...perimental psychology at... joined what was then called the Air Ministry as a ...chologist and stayed with it... mainly at UK Branches at ...Headquarters of Flying Training, Training and then Support Commands. Although most of his work on the training of pilots and of fighter controllers was classified, he has published in the *Air Force Quarterly*, and in the *RUSI* Journal, and has read papers at a variety of international conferences. He retired in 1982.

JOHN JAMES

The Paladins
A Social History of the RAF
up to the outbreak of World War II

Futura

A Futura Book

First published in Great Britain in 1990
by Macdonald & Co (Publishers) Ltd
London & Sydney
This edition published in 1991 by Futura Publications

Copyright © 1990 John James

The right of John James to be identified as author of
this work has been asserted by him in accordance with the
Copyright, Designs and Patents Act 1988.

All rights reserved.
No part of this publication may be reproduced,
stored in a retrieval system, or transmitted, in any
form or by any means without the prior
permission in writing of the publisher, nor be
otherwise circulated in any form of binding or
cover other than that in which it is published and
without a similar condition including this
condition being imposed on the subsequent purchaser.

ISBN 0 7088 4916 4

Printed and bound in Great Britain by
BPCC Hazell Books
Aylesbury, Bucks, England
Member of BPCC Ltd.

Futura Publications
A Division of
Macdonald & Co (Publishers) Ltd
Orbit House
1 New Fetter Lane
London EC4A 1AR

A member of Maxwell Macmillan Pergamon Publishing Corporation

Dedication

This partial and largely social account of the RAF in the period before the Second World War is dedicated to all the pilots and other military officers, both British and Foreign, and to all the scientists, who over the years told me what they knew about it, and even more important, what they did not know and wanted to know. And among those scientists, most particularly to my erstwhile colleagues Anne Brackley and Alex Cassie.

Contents

1 Introduction — 9
2 The Kitefliers — 20
3 The First War — 45
4 Transition — 69
5 Squadrons and Higher Powers — 92
6 Groundwork — 105
7 Worse Things Happen at Sea — 119
8 Officers and Pilots — 133
9 The Great Change — 153
10 Rebuilding — 167
11 New Foundations — 184
12 Plans and Pilots — 204
13 Consequences and Truth — 226
Tables — 239
Index — 267

Chapter 1

Introduction

On 14 August 1940 the German Air Fleets 2 and 3, aided by the presence of Air Fleet 5 on their right flank, began an attack on Southern Britain. This was met by 11 Group of Fighter Command, RAF, with 10 Group on its right flank and 12 Group on its left flank. Apart from seven weather days, fighting continued till 15 September 1940. After that date, the German Air Arm did not come again in force by day. In the old terms, they admitted defeat by leaving their opponents in possession of the field. The losses on either side in total or on any particular day are not important. There were no scores to be totted up: aviation is not a sport, it is war. What was important was that the Germans ceased to advance. They stopped for the first time in eight years, and went off, leaving the war against Britain unfinished, to fight a less formidable enemy in a more convenient place. This was the first battle between two air forces unaided and not fought as a subsidiary to any military or naval action. I have here tried to set out some of the events and processes which led up to this battle.

It is only fair for an author to say why he has written the book, although this is not done very often. It is especially necessary in a book on a subject which has been treated so many times already. The only acceptable reasons, which are usually taken for granted and are not stated, are that the author thinks that the books which have appeared so far have defects, and that he has material which has not appeared in print before.

What is more, none of the scholarly books, nor any of the popular works, told me what I wanted to know about the Battle of Britain or the Bomber Offensive. These were the affairs of pilots, men who saw aviation as a sport, not as war. They were the kind of men who shared in a mystique rather than simply a technique. They were, and still are, the men who have 'The Right Stuff', and who share the dreams of Jonathan Livingstone Seagull. And no writer told me what I wanted to know about them. There were detailed books but nothing in general, that is in numerical, terms. Nobody told me how many pilots there were in the RAF in the years before the war, how they were recruited, where and how they were trained, how they were organized, what they were paid, what they cost the country. And certainly there was nothing easily available which told me about the men on the ground, how they were recruited, how they were trained, what they were paid. I set out to try to write the book I wanted, avoiding the faults of earlier writers, and what I came out with was not quite what I expected; it was certainly not something which is commonly assumed today.

The Argument
The conclusion to which I was brought by actually looking at data is that the beginnings of British air power were due to the actions of a small number of officers, notably Colonel David Henderson and Captain Sykes, working as part of an organization which between 1905 and 1914 rebuilt the British Army and prepared it to go to war as an ally of the French against the German Army. Their original plan was for a single body, working for the War Office in a Military Wing, and for the Admiralty in a parallel Naval Wing. The Navy soon withdrew from the unitary system and went its own way. The Military Wing went to war in France as a Corps of the British Army, parallel to the scientific Corps of Royal Engineers or the Royal Artillery. In 1917, the then Prime Minister, Lloyd George, as a result of a piece of political fixing, authorized the setting up of a single service, the plan and the organization being supplied by General Henderson, by then Director of Military Aeronautics, on exactly his pattern of 1912.

After the war, there was no necessity, and probably no expectation, that the new service would continue. The Navy, who after the war were beginning to develop an entirely different use of air power in aircraft carriers, wished to withdraw at least the carrier pilots from the single service. The single service survived partly because the generals thought the Army was getting good value from the Independent Air Force, and shrank from the task of re-absorbing the RAF organizations and units into the Army. It also survived partly because none of his peers could cope with arguing with 'Boom' Trenchard in the relevant high level committees. During the 1920s Trenchard struck out boldly into creating an entirely new kind of military organization as far as the other ranks were concerned. The troops, a relatively intellectual group of technicians and technologists, resented being confused with the Army, but the officers at all ranks, virtually all of them pilots, seem to have continued their concept of the RAF as being something parallel to the Royal Artillery, and essentially a detached part of the Army for which it worked. During this decade, the number of squadrons in the RAF in the UK, and therefore the complexity of the organization, was greatly increased: the total number of pilots in squadrons, and therefore the fighting capacity, did not increase in proportion.

Not later than the summer of 1933, that is as soon as Hitler came to power, the Air Staff decided that they would fight the Germans not earlier than the spring of 1939. An expansion plan prepared independently of Army requirements was presented to the Government of the day and approved in November 1933. The plan, consisting of a Task Chart and a Manning Plan, was already necessarily complete in most details, and once the expansion had begun there was no realistic way of halting it or altering it in any substantial terms. The first element was a building plan, completed in the main before 1939, and intended to house a training system first for technical tradesmen and then for aircrew, and only then to house the operational squadrons. The expansion of the system of nodules in the communication and command network, that is the formation of new squadrons, was completed by the end of 1937. The existing squadrons were then stripped of their pilots to staff the flying training schools. The flow of new pilots into the squadrons did not begin till the middle of 1938. The tactics planned for the war were based on the air exercises of 1933, carried out for that purpose. The command system and the manning of the squadrons were completed exactly on time in 1939 for the traditional campaigning season of late summer.

This period of preparation marked the real separation of the RAF from the Army as a service able to wage war on its own account. This ought to be the end of the story. However, I have tried to show that the events of the first months of the war were used to test the tactical and strategic concepts of 1933, and that these were then radically altered, although the result was that the RAF became committed to a night bombing campaign using aircraft designed in 1933 to 1934 for day navigation and bombing.

Defects

The defects of the books which have already appeared on the Battle of Britain, on the lead-up to the battle and on the history of the RAF in general, fall into two categories, which may be summarized as romanticism and naïvety, ills to which professional historians are very subject. Usually they are combined.

Naïvety is seen in the over-enthusiastic use of verbal written evidence, the speeches and memoirs of politicians and the letters, and sometimes the memoirs, of officials and generals, the proceedings of committees and the minutes of meetings. The official documents are now for the most part open to general view in the Public Record Office at Kew. They are usually treated with considerable innocence and a touching belief in their virtue and truth. None of these documents, published or not, can be treated as a good source for our knowledge of what really happened.

Historians forget the words of a great figure in the English search for truth, Miss Mandy Rice-Davies. When told that a noble lord denied that he had slept with her, the lady replied, 'Well, he would, wouldn't he?' What she meant was that as a law of nature any man, called on to account for his actions, will so frame his words as to lead listeners or readers to put the best construction on them. This warns us that any autobiography or any politician's speeches must be read as an attempt to put the bravest face on things. Politicians' speeches are openly intended to influence opinion and should only be read with reference to the intended aim: this is well known.

The war memoirs of any politician or general or scientist or great administrator should always be read as an attempt by this man to say that he himself, by his unaided efforts, won the war, or at least that the war could not have been won without him. The further we get from the war in question, the more likely we are to find a minor figure making this claim, since any who could contradict him are now dead. As one reads more and more of these memoirs, it becomes obvious that each of the authors will avoid mentioning any other characters in the period, except where he can crow over them as defeated rivals and so exaggerate the importance of his own contribution.

Biographies of the great are even more suspect than their autobiographies. The trouble here is romanticism. It is very difficult to write a biography, a task usually occupying some years of work on dusty documents, without tending to make it into a novel and to treat the subject as your hero: if you don't do this, after a year or so you come near to dying of boredom and your writing shows it. If you have a hero you look around for a villain, and find one, however unjust you may be to the victim, although in somebody else's biography of him, he is, of course, a hero. Finding a villain in a military biography is easy, since all the men at a high rank may be seen and shown as competing with the hero for an even higher and more rarified rank; but ambition is a military virtue, and hard work to overpass one's contemporaries by fair means or foul is not a vice. The rules of British society lay down that a man must be moderate in his claims to greatness, if only for fear of contradiction and disproof and social discredit. But a biographer is under no such compulsion, and a biographer who works in close contact with a still living subject is most suspect of all.

Romanticism also blinds us to the fact that civilian virtues are not related to military ones, or to military efficiency. Modesty is not relevant. The fact that a man proclaims that he is the only one who can save the state does not necessarily mean that he is *not* the only man who can save the state. That their subject would himself lie, cheat and steal, and tolerate those who do, is a fact usually concealed by biographers intent on making heroes: nevertheless many rakes have been known as brave and honourable men in military dealings and as highly effective leaders and organizers in war.

Doctor Johnson remarked that every man thinks the less of himself for never having been a soldier. In this century the word pilot may be substituted for soldier. The romantic urge prompts every writer on the air aspects of this century to spread himself on the heroic deeds of individuals, and on the fights of the aces. Alternatively, there are a number of record-breaking exploits, especially those of the Long Range Development Unit which prepared the Bomber Offensive, and a unit of this kind must have a store of tales to be told. This book has avoided these attractive areas, because they have been well covered in the past and the space here was needed to tell a new story.

Romanticism also affects every effort to tell the history of the air arm. Writers make the service the hero, and feel obliged to show the whole sequence of human evolution from the amoeba as entirely directed towards the creation of the RAF from nothing, and by Trenchard. This is not necessarily true. Historians are not only romantics, they also tend to ignore the logical principle of Occam's razor. This is usually expressed in Latin or in English as 'Entities must not be multiplied beyond necessity,' but is more intelligbly expressed as 'Don't make things more complicated than they have to be.' The ambitious academic seeking at all costs to provide a new thesis topic is compelled to ignore this principle.

The effects of a process have nothing necessarily to do with the motives behind the operators. The results may be morally good, but the actors who produced them may have had very doubtful, or quite irrelevant motives. The RAF as an independent service is likely to have originated in a desperate attempt by a Prime Minister to save his own political skin since he believed that only his survival could rescue the country. It may not be dignified to say this, but it looks very probable: Trenchard's saving of the RAF by his own arguments and actions was dignified and well motivated enough. But there were heroes before Trenchard.

Historians therefore tend to be over-romantic. They also use bad sources. This is not merely a case of being taken in by great mens' memoirs. The memoirs are often useful in the details, in the same way as the memoirs of the little men are useful. It is usually said that generals write books on the art of war, or on the pleasures of poetry or painting, while air marshals write books about what fun flying was when they were young: admirals do not, or perhaps cannot, write. Historians imagine that by restricting themselves to the official papers from Kew, they are being objective, and dealing only with facts. This is naïve in the extreme.

Lord Armstrong remarked on a celebrated occasion that he was being economical with the truth. Those who burrow into the official papers must beware that the truth is being very economically doled out to them, and doled out by officials with axes to grind: two successive officials may grind very different axes on the same documents. The minutes of an official meeting at a working level in a service ministry are not always a true record of what was said. They begin as the notes made during the meeting by a Secretary who is not a shorthand writer but is a participating member of the committee. Draft minutes are circulated, and altered as members make plain, often on the telephone with no objective record, what it is they wished they had said, or what is their departmental line. The final *agreed* minutes go on to the files: the draft minutes are usually not preserved. Later, the files are deposited in the archives. Before they reach this stage, the original department will have done a fair amount of pruning. On deposition, the files will be pruned again in the interests of space. The resulting paper stock is suspect. The papers left *may* not be altered once deposited, but they will not reveal what has been extracted, or why things were extracted at different times, or what were the unwritten bases or assumptions or prejudices which the pruners, or the committee, or the writer and receiver of a letter, took for granted. The official papers are not a Good Source.

There is another patch of romanticism which dominates many books: nearly a hundred years ago Kipling pointed out that 'all unseen, Romance brought up the nine-fifteen'. This is the fascination with machinery and gadgets. There are enough books which trace in fine detail the development of the aeroplanes and of the radar system. In may also be remembered that Dr Johnson was dined in mess, lavishly, by the 37th Foot (North Hampshires) at the new Fort Augustus outside Inverness. Boswell remarked that he was struck with admiration on finding on this barren sandy spot such buildings, such dinner, such company. Dr Johnson replied more rationally that it did not strike *him* as anything extraordinary; because he knew that here was a large sum of money spent in building a fort; here was a regiment. If there had been less than what we found it would have surprised him. In the same way, we know how much money was voted to build an air force, and here still are the airfields. If there were not an efficient force in terms of machines and a command and control system of some kind in what was then the scientific capital of the world, then we should be surprised.

The Real Story

As I read through the literature of the years before the Battle of Britain, I became more and more doubtful of much of the material as based on 'Not Good' sources. In particular I became suspicious of so much romantic verbiage with never a single table of numbers. I went to look for a source of numbers. It is all very well to talk of God being on the side of the big battalions, but it is scholarly to say at least how big those battalions were. I found sources which were 'Good', and I describe them later on in the book. Setting out these sources led me to several clear conclusions.

The first was that to describe the growth of the RAF up to the opening of what we call the Second World War, I had to go back two or three generations before that happened. The growth of the RAF was not an isolated phenomenon, but was a feature of the general development of military and political history. Thomas Hobbes, besides saying that life in a state of nature was nasty, brutish, and short, also wrote, 'Warre consisteth not in Battell onely, or in the act of fighting, but in a Tract of Time wherein the Will to contend by Battell is sufficiently known, and therefore the notion of Time is to be considered in the Nature of Warre, as it is in the Nature of Weather.' In those terms, we must consider the first half of the twentieth century as one long European war; falling into two sub-periods. While the Second War was fresh in memory, it seemed so gigantic and cataclysmic in itself as to be a separate event, but as we move away in time, it is probable that our grandchildren will see the two wars as a single incident. The main difference between the two periods of hostilities in strategic and political terms was that in the interval Italy and Japan changed sides. Between these two phases, as it began to turn into something which could play a separate part in the action, the Air Arm on each side showed more and more consciousness that while a battle may be fought for control of the sea or command of the air, a war is fought for ownership of the land. And it could win that war.

The second conclusion was that at the beginning of its great battle, the RAF was not by any means as Terraine says in summing up the folklore picture, 'a small and inadequate force for the task it would be called on to perform'. On the contrary, it was designed for one specific battle in a particular confined area. The kind of battle it would be tactically, and the kind and number of units and of aeroplanes that would be necessary to fight it, were clear to the commanders seven years before it opened. The decision to fight once having been made by commanders grown old in wars and distrustful of allies, the building of a new air force was begun; once begun it had to go on, on exactly those lines then laid down. No political decision could alter what was being done.

The figures which led me to these conclusions are given in the Tables. This is largely material which I have not seen in print in this bulk before: as set out here, it is new. The raw data, untabulated, are however in print and open to anyone to read: in fact the inter-war governments *sold* the figures to the public. These Tables *are* the book. If you do not read the text but simply read and digest the Tables, then you will have done all that I ask of you. Figures are neutral. They have no emotional content other than the one we bring to them; they need only common sense to be understood.

However, the text does not only use the Tables, but gives the background against which they must be interpreted, a background not always familiar to those who read today about the Battle of Britain which their fathers fought, and even more obviously unfamiliar to many of those who write about it, either in histories or in fiction. I have found that the details of Royal Air Force history are something that we all think we know, but our picture, based on hearsay and gossip and vague memories, (or in the case of serving officers, on three or four hours of a rapid lecture course based on handouts founded in their turn on well censored and biased original documents) tends to be far divorced from facts. The facts are in the figures given here in the Tables.

I have found it convenient to bring the picture up only to the date of the Munich Crisis, September 1938. This is partly because the great question for anyone looking at the period is whether Britain could have gone to war at that time. My conclusion, based on the figures, is that we could have gone to war, but it would have been a most expensive war in casualties to a green and half-equipped Air Force, and one which we would have been very lucky to draw. And this limiting date is also set because after it my source of reliable data begins to dry up: whether because someone realized how much information was being given away, or whether because the effort of producing it for sale to the enemy was too great a strain on the clerical system, we cannot tell. However I have taken the story a little beyond Munich in examining the results, as late as 1943, of some of the decisions taken in the summer of 1933, as soon as Hitler came to power.

Once you have looked through these figures, no earlier book on the Air War will seem quite the same, or quite as convincing as it used to, and if you have plans to write, think them over first. Do read the text as well, though: you may find it worth while.

Notes and References

In a book like this where virtually every sentence can be tied to some written source or document, a full set of references would make a volume of unmanageable size, as would any attempt to review the relevant literature.

I have therefore kept references to a minimum, as far as possible restricting them to lesser-known books which it might not otherwise strike the casual reader in this field as worth seeking out. They all deserve reading.

It is more useful to cite books, which can always be obtained through the local public library, than periodicals, which can usually only be consulted by those having easy access to a copyright library. All these books (including the novels), form *part* of the essential reading list for anyone who wants to understand the military history of this century. Anyone wanting to follow the build-up to the Battle of Britain ought, if only for enjoyment, to read through the complete inter-war files of *Flight* and *The Aeroplane*, as well as of Brassey's *Naval Annual*, the *RUSI Journal*, the *RAF Quarterly*, and the *Journal of the Royal Aeronautical Society*. The *Air Estimates* with their accompanying *White Papers* and the *Air Force Lists* of the relevant years are such essential reading that I have not provided detailed references to them.

Any real history of a service must take as its model Professor Michael Lewis's two volumes on *The Social History of the Navy 1792-1860*. Of virtually equal value as guides to methodology are Brigadier Shelford Bidwell's *Gunners at War*, and Squadron Leader Neville Jones's *The Origins of Strategic Bombing*. Working through these will give the reader a new perspective on the military world and how to understand it. I have assumed that the audience will include not only that elusive figure the general reader, but also Battle of Britain buffs, military and political historians and other scholars, and officers and other ranks of all three services, but particularly of the Royal Air Force.

Chapter 2

The Kitefliers

The Old Army

In the summer of 1892, Hugh Trenchard, born in 1873, the son of a solicitor of doubtful financial stability, was commissioned into the Royal Scots Fusiliers, as a result of his success in his third attempt at the limited examination for militia officers who wished to enter the Army without preliminary training at the Royal Military College at Sandhurst. He had already failed the entrance examinations to the Royal Naval College at Dartmouth and to the Royal Military Academy at Woolwich. He was posted to the Second Battalion and travelled to India to join it.[1]

In that paragraph is summed up a series of massive revolutions in the methods of selecting and training officers for the regiments of the British Army. These revolutions, which bracketed the year of Trenchard's birth, were the work of Cardwell, the then Secretary of State for War, and are usually known by his name. Examining them and the subsequent reforms associated with the names of Haldane and Esher make clear many puzzling features of the modern RAF. It puts the development of the RAF in its proper place as part of the military and social history of the time, and it makes intriguing reading.

Before 1870, for many purposes the British Army was administered as two separate organizations under two different government departments and two Ministers. The 'Scientific Corps' of the Royal Regiment of Artillery and the Corps of Royal Engineers were under the authority of the Master-General of the

Ordnance and the Board of Ordnance; the Household Cavalry, the Foot Guards, the 28 Regiments of Cavalry of the Line and the 110 Regiments of Foot were under the authority of the Commander in Chief at the Horse Guards, and of the Secretary of State for War.

Since the seventeenth century the Board of Ordnance had been in charge of the construction of permanent fortifications and of the manufacture of cannon for both the Army and the Navy. Its main requirement of its officers was professional skill and competence. The young gentleman who wished to enter this service as a career must obtain a nomination by the usual methods of social and political influence and then be examined as to his academic suitability. At first the examination was an assessment of attainment before entering on a personally tailored course of indeterminate length, the entrant being allowed out on the world when he was considered fit for service. Later it turned first into a qualifying and then into a competitive examination as a prelude to a very formal and rigorous course of training. Both Corps began to go into the field with the Army during the eighteenth century, first as technical advisers and later as combatants.

Once the young man had been commissioned, into one or other corps, he remained in that corps for the remainder of his career, until he became a colonel, after which he might be selected for a General Officer post and then command a force of all arms. His promotion in his corps depended entirely on his experience, that is on his place in the seniority *List* of officers.[2]

In contrast each regiment had originally been raised in a system of privatization, by a military entrepreneur who also acted as commander, the colonel. His motive was to make money from his role as the agent who bought for the regiment its clothing, food, equipment, arms, ammunition and doled out its pay; he claimed afterwards from the Crown, making what profit he could on this deal. Unsympathetic reformers slowly chiselled away at the colonel's sources of income until the 1850s, at the very end of the system, only the profits on the clothing contracts were left on which it was officially calculated each colonel made about £6,000 a year.[3] Command of a regiment was the peak of every officer's ambition: naturally.

The main concern of the regiments was political stability. The officers were therefore drawn from the aristocracy and the landed gentry, and later the wealthy merchants and even manufacturers. The young man who wished to be an officer in a regiment, either as a career or till he found something better to do, would (or his relations would) find a post at the lowest level, as a Cornet of horse or as an Ensign of foot, where the present occupant wanted to move, either out or up, and pay him the regulation price for his post. Most civil service posts were held on the same terms till reforms started in the first quarter of the nineteenth century. Promotion came in the same way; the cornet looked for a lieutenant moving out or up, and paid him the regulation price. Approval had to be sought for each move from the colonel of the receiving regiment, and from the Horse Guards, and there were complex regulations governing speed of movement and the price: corruption was possible and attractive, and one mistress of a Commander-in-Chief made such a good thing out of fixing purchases that even Regency society was scandalized. The basic fact was that an officer's promotion, up to the rank of colonel, was dependent not on his valour or his military efficiency, although a conspicuous lack of either might make a colonel think again about approving a purchase, but on his skill in financial wheeling and dealing, and on shrewd movements between regiments, even between foot and horse and back. A colonel was eligible for promotion, through the usual channels of political and social influence, to a General Officer post.

Cardwell was the latest in a series of reformers. One of his forerunners had been Sweet William Duke of Cumberland, the 'Conquering Hero' of the song and the saviour of his country.'[4] The victor of Culloden does not usually appear as an enlightended reformer, but he fought for years for the abolition of purchase and the assimilation of officer entry and training for the regiments into the system in use for the Scientific Corps. That was, in essence, what Cardwell brought off. Cumberland had been defeated by the landed gentry and the aristocracy who saw entry to the officer ranks as a prerogative of their own sons seeking temporary employment or a career. By the time of Cardwell these groups were no longer the dominant class politically but they could still give trouble. Therefore, these landowners, who controlled the Militia and whose sons made up the officer ranks in it, were allowed to send them into the Army without the expense of training at Sandhurst, after a qualifying examination and experience in a militia regiment. Between 1870 and 1914, Sandhurst accounted only for about two-thirds of the officer entry to the regiments. Trenchard took advantage of this loop-hole.

The other major reform came into effect from 1880, and affected only the old numbered regiments of foot. The twenty-five lowest numbered regiments were each already organized as two battalions. Now the higher numbered regiments were paired up as two battalions of new regiments. Each new regiment was given a recruiting area, a depot within that area, control of the militia regiments in that area, and a new title derived from the area. For instance, the 30th (Cambridgeshire) Foot and the 59th (2nd Northampton) Foot were united as the East Lancashire Regiment.

Thus were born the County Regiments, the basis of the main myth of the British Army. In this myth, every soldier, officer or man, enters a regiment of which he remains a member all his military career. It is his military family on which he is expected to centre all his loyalties and emotions, and which he expects to look after him in sickness and in health. The physical symbols of this loyalty were the Colours and the cap-badge.[5] It was the County Regiments which fought the three biggest wars of the country's history, which must be discussed here, and a vast proportion of the male population served in them; therefore it is difficult to realize that the system lasted only for about eighty years until they were absorbed, in fact if not always in name, into the big divisions of the 1960s.

A feature of the Cardwell scheme was that while the Depot looked after recruiting, recruit training, documentation and minor administration, one battalion of a regiment should always be in the United Kingdom – Ireland was a very useful place for parking the Army – while the other battalion should be abroad. Therefore the newly commissioned Second Lieutenant (no longer Ensign) Trenchard was sent out to India to join the Second Battalion. India was a cheap place to live, attractive therefore because as a relic of the old system the officer was still really considered as an amateur, a gentleman serving the sovereign for his own pleasure, and supplying for the purpose his own uniforms, personal weapons, and equipment, all to the official scale; and otherwise paying his own way, although being compensated for out-of-pocket expenses. Trenchard lived the normal life of any junior officer with no private income, making a little on the side in the gentlemanly pursuit of training and dealing in polo ponies. His service in India overlapped with that of another polo-playing junior officer, Winston Churchill, and it would be interesting to know if they ever met at this period; they must have heard of each other in that small closed society.

One effect of the Cardwell reforms was that officers no longer moved about from regiment to regiment, but now spent their careers between the two battalions; their men, or the other ranks as they were patronizingly called, had always remained in the same regiment into which they were enlisted for life, that is till they were killed or died of disease or were cast off, by reason of age or feebleness, to beg in the streets. When the First Battalion came out to India and the Second Battalion went home, Trenchard transferred. So he was in the First Battalion when the South African War broke out.

The South African War
The Second Boer War opened in October 1899. The armies of the two Boer republics of the Transvaal and the Orange Free State invaded the British Crown Colonies of Natal and the Cape. In both areas the British were soundly beaten. The senior British commander, Buller, was in Natal.[6]

By February 1900 the British Government had sent out reinforcements, an ageing but energetic Commander-in-Chief in Lord Roberts, and an efficient chief of staff in Lord Kitchener. General French was in command of the cavalry with Lieutenant-Colonel Haig as his Chief of Staff. Among others present on the staff were Captain and Brevet-Major William Robertson of the 3rd Dragoon Guards, Major and Brevet-Lieutenant-Colonel Henry Wilson of the Rifle Brigade and Lieutenant-Colonel David Henderson of the Argyll and Sutherland Highlanders. There was a corporal of Royal Engineers named H.H. Kirby, and Frederick Sykes came too. General U.S. Grant remarked of the Mexican War of 1848 that it was not very important or spectacular in itself: what was important was that all the officers of the regular US Army took part and there he got to know almost all the men he would fight with or against later when they were all generals.[7] In the same way, the Boer War is important for us because almost everybody who took an important part in the foundation of British air power was concerned in it in some way, even Lloyd George and Jan Christian Smuts.

By May 1900, Roberts had invaded the republics, Kitchener had won a battle at the Modder River with a massive barrage directed by spotters in balloons, and French had led a vast cavalry sweep to Kimberley and north into the Transvaal. The organized Boer armies were destroyed, the two governments were fugitives, and it seemed the war was over. The Boers then resorted to guerrilla warfare and this broken-backed phase dragged on till May 1902. Roberts went home and Kitchener was left to finish the war; by the end he was using an army of nearly a quarter of a million men in South Africa, which with backing and rotations meant that half a million had passed through the ranks. Of these fifty thousand were raised by the government of Cape Colony and large organized contingents were sent by the governments of Canada, Australia and New Zealand. The guerrilla war proved immensely expensive in money, men and horses. The Colonial troops, British infantry and artillery, and large contingents of Militia and Volunteers were used as mounted infantry. There appeared to be no shortage of men, but officers capable of managing small detached bodies of men well away from their parent units and also leading them in action were scarce. Captain Trenchard, the trainer of polo ponies, found himself training and commanding bodies of mounted infantry, and in October 1900 he was shot through the chest and invalided home for treatment. He went to Davos to convalesce.

Frederick Sykes was a few years younger than Trenchard – he does not give his date of birth in *Who's Who*.[8] His origins are obscure, and in contrast to his early years, Trenchard's early career, as narrated by Boyle, seems sheltered in the extreme. Sykes was brought up in Kent. By the time he was sixteen he was working in Paris as a clerk in a retail business. At eighteen he was back in London as a clerk to a firm of merchants in the City. The next year he went to Ceylon to be assistant manager of a tea plantation. After eighteen months he set off for home again, by way of Japan and America, travelling as he could, working here and there. He reached London just in time for the start of the Boer War, and decided to join in. He bought himself a rifle and a colonial saddle, and set off for the Cape. He joined one of the units of irregular cavalry raised by the Cape government, and rode and fought across most of the colonies and the republics. He was captured by the Boers, force-marched from camp to camp one jump ahead of the pursuing British, was abandoned, picked up, returned to his unit, and commissioned. He was shot through the chest, invalided home to Britain, went to Davos to convalesce: it would be tempting to think he met Trenchard there, but the dates do not fit. Sykes returned to his unit in South Africa.

When the war ended, Sykes was offered, and accepted, a regular commission in the 15th, The King's Hussars. He does not tell us how this was arranged, but someone must have noticed him. The Hussars were in India, and Sykes went there to join them. Immediately he began to look for action. There was a call for junior officers to volunteer for service in a small war preparing in Calabar, in south-east Nigeria. Sykes persuaded two other subalterns, friends, to come with him: this is typical, Sykes always seemed to have friends while Trenchard had only followers. Sykes reached Calabar to join a pool of junior officers on the coast waiting to fill up vacancies caused by casualties, which were expected to be heavy. What Sykes said in later years when he found that he had volunteered to serve under Trenchard is not recorded.

Captain and Brevet-Major Trenchard (that is to say, he had been promoted so that he was a major on the *Army List*, and was paid as a major, but he was still a captain on his *Regimental List* and would have to wait his turn till he became a major on that and eligible to command a company) had volunteered for service as second in command of the force in Calabar, and was in command when it moved up country. He was an economical commander, and the expectant junior officers in the replacement pool were not needed. Sykes went back to India by way of accumulated leave in London. He spent it going on courses.

For Frederick Sykes, who had been on his own since he was sixteen and had made his way around the world, was convinced that his comrades were better officers than he was because they had been trained at Sandhurst and therefore knew more than he did. The way to learn more was to go on courses: drill courses, musketry courses, signalling courses, hygiene courses, machine-gun courses, everything. By the time he got back to his regiment in India he was convinced that there was one essential course he must go on to be a proper well trained soldier. He must go to Staff College. He found himself a posting to the Intelligence Staff at Simla, where the example was always before his eyes. William Robertson had enlisted as a trooper at the age of 18 in 1877 and by the time he was 25 he was a regimental sergeant major. He was so impressive a man that a meeting of the colonels of the cavalry regiments in the United Kingdom was called to discuss ways of commissioning Robertson in a regiment which was about to go to India for a long time, where he could afford to live on his pay. After seven years in India, Robertson made an advantageous marriage, and was able to afford the expense of going to Staff College. In another seven years he had passed from the oldest lieutenant on the *Army List* to the youngest Lieutenant-Colonel. He was now a Brigadier-General on the new General Staff, and rumour was that he would go even higher.

The General Staff
The Great General Staff was a Prussian invention, to fill a specific military need. Prussia was a flat country with no natural boundaries, open to any invader, like Napoleon who had destroyed the Prussian Army at Jena in 1806. The only protection was a better trained and organized army. The old Prussian Army, raised like the British on a system of proprietary regiments, would not do. The generals after Waterloo began to set up a better system. The Commander-in-Chief was given a staff, organized as a Staff Corps, with a distinctive red stripe down the trouser leg. The Staff Corps was recruited by selection of the brainiest officers from each year's intake. They were sent to a staff college where they were given an intensive course on the principles and the detailed carrying on of war.[9] The best of the output were posted to the Staff Corps and employed in Berlin. Here the General Staff planned the next war, against one of the country's neighbours. The Army was manned by conscription and a reserve was created. The regiments were grouped permanently into brigades, divisions and corps, and the headquarters for these higher formations existed as permanent formations.

On the outbreak of war, the reserves were to be mobilized and the regiments sent off to the threatened frontier. The marches had already been planned so that the regiments on their way did not collide, jostle each other, or try to occupy the same stretch of road at the same time, and so that each regiment at the end of the planned day's march would find a place to spend the night with food ready. The regiments, arranged in their brigades and divisions and corps, would thus arrive at the proper places on the frontier in the planned formations and ready for action. All this planning was the business of the General Staff. When railways were invented the task became slightly easier because the job of the staff was now the production and updating of complex railway timetables.

There were a few time conditions. Proverbially Kings in Europe do not march to war till after the barley harvest, that is late July. The result of the General Staff's calculations would be one decisive battle, either on the frontier or, preferably, just across it to reduce damage to the home country. One battle would be enough, and then the Army could go home by Christmas.

Three generations after Napoleon, the Prussian General Staff were ready for what were, essentially, a series of preventive wars to establish the physical frontiers. The first, in 1864 against Denmark, was hardly a contest. It gave Prussia the space to dig the Kiel Canal. In 1866 came the war against the Austro-Hungarian Empire. This started a little early, in June, because most of the other German states sided with Austria but not to the extent of joining in yet; action was necessary at once to forestall them. The Austrians were knocked out at Königgratz, and the Prussian armies were home for Christmas.

The next was the big one, against the French. The Prussians, allied with most of the other German states, invaded at the beginning of August. Thanks to the General Staff the Prussians concentrated more efficiently and the French field armies were destroyed at Metz and Sedan and the Emperor captured by the end of the month. The armies ought to have been home for Christmas. However the French refused to give in, and provoked the German armies into a siege of Paris. This lasted till the end of January, while rather ramshackle armies were raised all over France; the German lines of supply were put under pressure, and the surrender of Paris was for the General Staff rather a matter of relief than one of rejoicing.

The German General Staff (no longer simply the Prussian) went back to their drawing boards, while the other governments got out theirs. They were deeply impressed, and the most usual response was to ensure future success by adopting the Prussian helmet as a headgear for their armies.[10] The French did not. Military thinkers and writers in Britain decided that the General Staff had been responsible, and advocated the setting up of a staff for the British Army on the Prussian model, supplied by a staff college, on the Prussian model also. The usual emotive term was 'the brains of an army'[11] and this is probably the first time that anyone had suggested that the British Army needed any directing brain: political stability had so far been enough.

What the British Army did not need however was a general staff on the Prussian model, because it did not intend to fight a major war on the frontiers as the Prussians did. It fought wars of a quite different kind, and it had long had a staff college well adapted to its own peculiar needs.[12] The type situation of the British Army had been seen, on a large scale, in the South African War, especially under Buller in Natal. The usual setting was that of a colony or province where the Governor found himself coping with a sudden emergency, an outside invasion by the Shah or the Czar or some such potentate, or a rising of the local aristocracy or the kings of the hill country. The Governor might have, say, three battalions, and half a regiment of cavalry, usually locally raised with some British stiffening and British artillery. They would be scattered in stations convenient for housing troops but not always convenient for concentration.

The Governor would have no military staff, only a couple of aides-de-camp chosen for their social and diplomatic skills and family connections rather than for their military expertise; in other words chosen for the work they would spend most of their time doing. What the Governor needed at once was a handful of officers who could put a campaign together for him, who could work out on the scanty maps available (making their own where necessary) how to bring the scattered troops together without any traffic jams so that they could all be arranged in the right order at the concentration point at the right time on the right day. It was easy enough to recognize the need for transport carts and food: the clever thing was to work out how much food and how many carts and where and when, and how to cater for the irreconcilable needs of rice-eating regiments and bully-beef-eating regiments.

Once a well-fed army could be got together a battle could be arranged. What the Governor, now turned into a Commander-in-Chief, had was a handful of young officers, good riders, who could take his orders to the commanders of regiments, and see that they were accepted and obeyed: their junior rank was balanced by their being known to have been in on the planning and knowing what the general was thinking. In such a situation what was needed was not a general staff in London but a general's staff on the spot.

The Staff College at Camberley had been set up for just this purpose, to produce throughout the army a sprinkling of men qualified to join, at short notice, a general's staff. The course at Camberley taught them how to lay out marches and routes, how to calculate food and transport needs, how to draw maps which other people could read, and how to ride: carrying orders fast over rough country under fire would be an essential part of the staff officer's work. Of the teaching methods, the most famous were the Staff Rides, exercises in going across country and solving in the field the problems set not so much by the staff as by the terrain itself. In contrast the German General Staff worked, and trained, by solving war games indoors and under cover.

So affairs went on till the Boer War showed that the County Regiments worked and fought very well, while the staff work was dreadful not only during the first emergency period of Buller's defeats but also later during the formal warfare and the broken-backed phase. Intelligence was not gathered, reconnaissances were not made, troops attacked across unknown country against unlocated enemies while uncertain where their neighbouring units were or where their food and water and ammunition were coming from, artillery batteries were galloped into action half a mile ahead of the infantry they were supposed to support. Above all, neither battles nor campaigns were planned but developed haphazardly. All this *was* the fault of the staff: but there weren't enough staff-trained officers, and amateurs had to do the work. Few generals were staff trained since uninterrupted regimental service was the way to promotion, and they didn't take their staffs' advice seriously: the staff lost control of their generals, and how else do you think a whole army got besieged in Ladysmith?

After the débâcles of the Crimean War it took sixteen years to set up the Cardwell reforms. The repairs to the structure after the end of the Boer War started within two years. The Army had learnt some things from the conflict. It was comparatively easy to raise large numbers of men, but the physical condition of the nation was dreadful and at least half of the volunteers were unfit for immediate service. In any case they were untrained, and even militia were better than raw recruits. The county regiments were efficient at calling out volunteers by way of their higher numbered battalions which had once been militia regiments. But the battalions were not easy to handle unless they were grouped in brigades and divisions, and they needed practice in working together in these new formations. Permanent formation headquarters had to be established and made ready to go into the field. And in any case neither battalions nor higher formations could be trained properly unless there were some indications as to what they were being trained for.

As to what they were to be trained for – that was something to think about. The times were changing, and Germany was becoming dominant in Europe. The French were quite sure that they would have to fight sooner or later, and were looking for allies. The King went to Paris, and by 1904 there was a loose agreement between the two governments; not a treaty, but an understanding, and enough for staff talks to be based on. If there were a British staff.

Esher

There was going to be a British staff, set upon the German model, to plan one war, one battle. In the usual way the British Government appointed a Royal Commission to consider the state of the Army. The chairman of the committee was Reginald Brett, second Viscount Esher.[13] Esher was what one must call a professional courtier, a favourite of Queen Victoria and of Edward VII. He knew everybody, influenced everybody, kept out of the limelight and even after succeeding to a peerage always refused any office which might make him too obviously attached to one political party. He was given a high civil service appointment as Secretary to the Office of Works. He seems to have treated this as a sinecure, his real life's work being in politics.

When the Royal Commission was appointed, the Conservatives were in power, but in 1905 they were replaced by the Liberals led by Campbell-Bannerman, succeeded by Asquith in 1908. One of the leading members of the Liberal administration was the noted pro-Boer pacifist David Lloyd George.

Defence was only one of the problems with which politicians were concerned, others being trade union legislation, old age pensions, National Insurance, education, the legal position of the Church of England in outlying areas like Wales, and unrest in Ireland. The world was very different then. The upshot was that Esher was able to work without very detailed interference. What he did was at first over the head of the Conservative War Minister, Arnold Foster, and then with the enthusiastic cooperation of the Liberal War Minister, Lord Haldane, with whose name the reforms initiated by the Esher committee are usually associated.

The Esher reforms began at the top with the creation of a Committee of Imperial Defence, with the Prime Minister as Chairman, and Esher among the permanent members; he left the Office of Works. He thus retained his influence on military matters for a long time to come, much of the later work being done through this committee and its sub-committees. At the basic level the Army was given a formal organization into brigades, divisions and army corps, with permanent Headquarters at all levels and an overall structure designed for immediate participation in a major war on the mainland of Europe. The Commander-in-Chief was replaced for most functions by a Chief of the General Staff – the Imperial General Staff later, hence the CIGS – with a planning staff on the German model subordinate to him. Other measures were taken to create reserves which will be discussed in due course; an important step was to create an appointments board but to reserve approval of general officer appointments and promotions to the King.

There was no real doubt in the minds of anyone who thought about it that the enemy in the major continental war for which the Army was now being prepared would be the Germans. The staff talks with the French could now begin. The new General Staff could fulfil one of the main elements of the definition of the General Staff; it was designed to plan one particular war, perhaps just one particular battle.

This then was the staff world which Sykes intended to enter, by way of the Staff College with its revamped course. It is likely that he understood what was going on in politics, like any other officer who worked at his profession, read French and German and wished to get on and survive. In those days an officer who wished to go to the Staff College had first to pass the qualifying examination (known as the 'Q'), for which study under a specialized crammer was considered essential, and which was held only in London. A candidate had to pay his own way to London, even if he were in India, and back, and while he was on the course he was put on half-pay. This financial penalty operated as a kind of selection procedure, as only the man who was really keen would put himself down for the examination. We may compare the one good feature of purchase, which ensured that if a regiment went overseas the officers who went would be either those who could not afford to buy a step up into another regiment or exchange out, or those who were really keen on the exercise of the military profession; for either category there was a better chance of free promotion into a death vacancy, since if a regiment were in, say, a yellow fever area no one was likely to want to transfer from elsewhere into the death vacancy.

Going to the Staff College was expensive enough to deter anybody, but Sykes, a bachelor, wanted to go. He came home from India, went to a crammers, sat the 'Q' and waited for the results. And to pass the time while he waited for the results, he went on courses. In particular he went on the ballooning course.

Balloons
Ballooning is now a sport, but there, so is rifle shooting. Ballooning then was a well established part of the Army's system. The balloons had come into the army in 1873, when the men who had started experimenting after hearing of the use of observation balloons in the American Civil War were able to point to the use of free balloons for travel out of Paris during the siege. Balloons, and the Balloon School and Factory, belonged to the Royal Engineers, who operated them at Farnborough.[14]

Ballooning was an art, a military skill which had to be learnt. The Army wanted balloons in order to do what every commander wants done, that is to see the other side of the hill, to see what the enemy is doing and to direct gunfire on the unseen targets.

Aerial reconnaissance became necessary at this time because there had been a visibility revolution in war. Armies of earlier times had worn bright uniforms so that the enemy could find them when they went out in masses on perhaps a front of a mile to 'offer battle'. But the effect of the American Civil War was to make it plain that battles were possible under telegraphic control over fronts of eight or ten miles. In the 1880s smokeless powder came in and armies began to wear camouflage uniforms, khaki or horizon blue or field grey. Experience against the Boers had taught respect for the 'Mauser's 'arf-mile 'ail'. Reconnaissance by cavalry troopers without military sophistication was not always successful. If responsible officers could get into the air and look down on the field, they could find the enemy.

It wasn't as easy as that. From five or six hundred feet, the world didn't look like a map, the rain and fog got in the way. The observer had to learn to spot what he was looking for, and how to report it. And he had to do his looking and spotting under dreadful conditions. He had to make his notes or at least unfold his maps with frozen or heavily gloved hands. The balloon didn't keep still. A round balloon spins and tosses about in the turbulence, and airsickness is not unknown.

By 1905 it was known that the Germans had thought of a way to stop the balloon tossing and spinning, by making a sausage-shaped balloon instead of a round one, and by fitting horizontal and vertical stabilizers at what became by courtesy the back end. This was meant to keep the balloon's nose into the wind and stop the pitching. Seeing what ought to be done is easier than doing it oneself. The Sappers tried but found it very difficult to make and operate, and the Army went to war in 1914 with round balloons. By 1916 they had adopted a French technique of making stabilizers and this worked. The balloon with stabilizers was confusingly called a kite balloon by the British and a *drachen* by the Germans.

The Balloon Battalion was very much a mounted, or at least a horsedrawn, affair. The technique was to select a field which was large enough and flat enough. The wagons arrived, and the Sappers laid out the envelope and sorted out the netting and the basket, just as one does today. When everything was absolutely ready, including the observers, the envelope was filled with hydrogen from cylinders and the balloon went up. The admirals and generals were obsessed with the vulnerability of balloons, and their relations the airships, to artillery fire. There was a fond delusion that these big gasbags would be easy to hit with ordinary field guns. It took the lessons of war to show that the only way of knocking one out was to fly as close as you could in a heavier-than-air machine and shoot at it with incendiary projectiles in the hope that it would catch fire: then you had to dive and get away as fast as you could in case you had set it on fire since it just might explode. It was not easy to hit a balloon from an aeroplane if it were in turbulence and the ground crew were trying to haul it down as fast as they could. And an airship might easily be carrying out evolutions of its own to make you miss, and shooting at you with its own guns. In the early days of the war the approved technique was to fly over the airship and drop bombs on it, presumably with impact fuses.

The balloon was vulnerable to artillery but in a different way. The moment the balloon rose above the skyline, every enemy battery in range would see it, and start ranging and calling for approval from brigade for shifting target. The desperate need was to send the balloon up as quickly as you could, make the observations, haul it down, deflate it, pack envelope and basket into the wagon and *gallop* out of the field before the shells started to arrive into what the enemy hoped would be a profitable assortment of balloon and men and wagons and horses and hydrogen cylinders. Speed was essential.

A balloon could not be operated in bad weather, certainly not in winds above a moderate speed. But the mention of the kite balloon sparked an idea in many minds. The War Office engaged an American inventor, a Mr S.F. Cody, who was trying to perfect a man-carrying kite. This seemed quite feasible, and it was. The kite could be launched in moderate winds and stronger, and if there weren't a wind you could get lift by hitching the kite behind a vehicle, horsedrawn in those days. By 1905 Lieutenant Broke-Smith had set the height record for a man carried by a kite at 3,340 feet, and Colonel John Capper, then Commandant of the Balloon School, had been carried for several miles during manoeuvres, 'whilst he observed and reported on the course of the battle and the enemy positions and movements'.

Colonel Capper is remembered usually as the man who turned down the Wright brothers. This is not fair. In December 1904 he went to the Saint Louis World Fair on behalf of the War Office to see, bluntly, what he could pick up of any use. He heard of the Wright brothers, and went off to Kitty Hawk to see them. The season was over, so he had to go back after them to Dayton. They were willing to talk, and assured him they had, really, flown a heavier-than-air machine. They had two photographs in which the eye of faith could see that the wheels were actually off the ground. They would come and work for the British Government. Their needs were simple. They only wanted £20,000 down, permanent and well paid appointments, a vast open space for their experiments, and a factory with ample manpower and materials. And if they found anything, they would let the British Government, who were paying for it, see it just as soon as they had told the US Government. Capper was in fact very impressed, and recommended that the War Office should take up the offer. The War Office decided that they wouldn't.[15]

This was not unreasonable. The evidence for flight was very thin, and not enough to impress the US War Department who were, after all, nearer and better able to monitor the project. The steps of every War Office in Europe were crowded with inventors who were sure they had found the secret of flight and only needed a down payment, a good regular salary and massive support in men and materials. At least they were not usually insistent on giving the secret, when they found it, to other governments first. It seemed wiser to wait, and in another year the secret would be out, the technical grape-vine being what it was.

In another year the secret *was* out. The Wrights' machine was not at all what the generals wanted. They wanted a carriage in which the observer could float above the battle and look down in comfort while writing his reports. Something like this was already becoming available in the airships of Count Zeppelin. But they were enormously expensive, and very cumbersome, needing vast numbers of men to move them about when they were at base, and unable to operate in really bad weather. The Wright brothers had abandoned the idea of a stable machine, and had gone for an essentially unstable vehicle, in which a man sat in the middle and kept the thing in balance by constant corrections. If we were to invent it from scratch now we would, of course, install a computer to carry out the corrections and feed it data from a system of gyros, but computers had not been invented at that time: a man was easier. But a heavier-than-air machine was expected to be cheaper than a Zeppelin, and to need less space and to fly in worse weather – little did the generals know. The War Office already had its own American engineer and inventor, paid a regular salary and with a factory and ample men and materials at his disposal. He had succeeded in solving the man-carrying kite problem and he could now be given the powered flight problem. He could carry out his experiments on Laffan's Plain when it was not in use by its regular occupants, Royal Engineer field companies.

Mr Cody was happy to oblige, and at the same time the Royal Engineers were given the task of building an airship. There was only one engine available for them – not one type, but one individual engine which they had to share, and it could not be fitted in two machines at once. But there was a high degree of co-operation, and in October 1907 Dirigible No. 1 flew from Farnborough to Saint Paul's Cathedral, but could not make it back against the wind, coming down at Crystal Palace. Colonel Capper was the pilot, and Mr Cody was in charge of the engine.

Staff Duties
This was the world to which Sykes' balloon course introduced him. Alas, he failed the Staff College qualifiying examination, and he went back to his regiment, now in South Africa. There he used one of his courses when he was ordered to set up a machine-gun school. Meanwhile, another Staff College had been set up at Quetta: very reasonably, since about half the British Army was in India and there was also the Indian Army to be catered for. Sykes sat the 'Q' again, and passed. So he went to Staff College, but in India, not at Camberley. The Commandant of the Staff College at Quetta was Major-General Thompson Capper, the brother of the Commandant of the Balloon School, a dynamic man and an effective teacher. The burden of his teaching was that it was the duty of a staff officer to be killed in action.

When the course was over, Sykes took accumulated leave to go back to London specifically in order to learn to fly. He had also arranged to spend a month in Germany for language practice. But almost as soon as he landed he was summoned to the War

Office to see Brigadier-General Henry Wilson, who had finished a tour as Commandant of the Staff College at Camberley, where he had been succeeded by Brigadier-General William Robertson. Sykes had been recommended to Wilson, presumably by one or other of the Capper family. Wilson, now Director of Military Operations, offered Sykes a job in Intelligence, which was under his direction, as soon as his German was good enough. So Sykes stayed in London and in the mornings of the autumn of 1910 went down to Brooklands to learn to fly while the air was calm. Also learning to fly at that time was Brigadier-General David Henderson, Director of Military Training. Henderson, born in 1862, must have felt himself a late developer beside young whizz kids like Robertson and Wilson, since they had all been Lieutenant Colonels in South Africa together, and he was ten years older than they were. Now all the threads began to draw together.

On 18 December 1911, a Technical Sub-Committee of the Committee of Imperial Defence was set up to consider the details to give effect to a proposed policy for establishing an Aviation Service. As often, the answer is in the question (what is called the hidden agenda), and we do not know who exactly asked the question. The sub-committee included Henderson, with Sykes as Secretary. The scheme the sub-committee produced was ready for approval by 25 April 1912. The main outlines of the scheme were the setting up of a single organization to be called the Royal Flying Corps. It was to consist of four main elements, a Military Wing, a Naval Wing, a school for training pilots and a factory for building aeroplanes. A reserve was also to be created.

The detailed organization of the Military Wing was left to Henderson and to Sykes. They decided to set it up basically as a system of small units. Sykes went to France and saw what the proposed system was there. The French were going for small units, perhaps half a dozen aeroplanes. The French might have their own reasons, but the British staff had been through the final phase of the Boer War, and saw the problems of small independent units, problems of supply and manning as well as operations. The basic unit for the Military Wing was to be the Squadron. This would be a major's command, because it ought to be possible to find enough majors to command squadrons: a major could be depended on to have some maturity and experience in doing the business of a small unit. A battery of field artillery was comparable as a major's command, with about two hundred men. A flying unit of this size would have a strength of twelve aeroplanes, with a stock of 'remounts', and its own administration, catering, transport and so on.

Sykes went into the details. He was seeing a new regiment, even a new kind of regiment, being created. One essential was the uniform and with it the cap badge. Sykes tells us that it was he who devised the 'Maternity Jacket', a double-breasted wrap-over tunic of, as he saw it, severely functional design, and he chose a colour for it, a shade known in the trade as French Grey. He also adopted the flat-side-hat. This may be called a forage cap, but the term is vague: in Victorian literature, the 'forage cap' means any cap authorized to be worn when the standard headgear, the brass helmet, say, or the busby or the lancer cap, is inappropriate. Often the term is applied in the Crimean period to the flat peaked cap as worn today by the RAF, but the side hat was certainly in use then. Sykes saw its advantages as a mandatory, not an optional, item of attire, and wrote it into the scheme.

Sykes, in his memoirs, is never grudging in attributing to the originators other innovations which may have passed through his hands but for which he was not otherwise responsible. When he claims credit for an innovation we may therefore take him at his word unless we have much better evidence. Thus he sent on others' designs for the cap badge and the flying badge, with full attributions. He also sent up the motto for approval. He acknowledges that Per Ardua ad Astra may be bad Latin, but points out that his education did not include luxuries like Latin, and the motto is not his invention.

The question of equipment then arose. Immediately there was no shortage of inventors ready to sell their aeroplanes to the Army. The Royal Flying Corps therefore held trials at their factory at Farnborough to select a standard aircraft for the Corps, much as the Army had a standard rifle or a standard design of wagon. Mr Cody entered his design which had been produced as a War Office requirement, but it was not thought suitable, for entirely sufficient reasons, like structural weakness. Two designs produced by the Royal Aircraft Factory, as the Balloon Factory had become, were adopted.

The Duke of Westminster had presented a Voisin pusher biplane to the Army, which had damaged it and sent it for repair to the Royal Aircraft Factory. Probably the most useful thing about it was the engine. When the aeroplane re-emerged from the Factory, or at least when that rudder number emerged, it appeared as what we would now call a conventional tractor aircraft, similar to those designed by Blériot, with the engine and propeller at the front. The reason is simple: the Factory had no funds for the design and construction of prototypes but there was money for the repair of broken aeroplanes. The new aircraft was called the BE1, or Blériot Experimental. A second version was built to comply with the specifications for the Military Aeroplane Competition, and although it was not formally entered it was superior in performance to any of the competitors. The subsequent versions of this BE2 were given distinguishing letters, and the BE2c was the standard equipment in 1914. It was designed for reconnaissance, as far as it was designed to do anything but simply get off the ground and fly, a difficult enough requirement at that period. It was further developed to carry bombs up to 500 lb in weight, or a wireless telegraphy set, and this ended as the famous RE8.

In 1910, a Mr Geoffrey de Havilland had been appointed Assistant Designer at the Royal Aircraft Factory. He had already started by designing and building a pusher biplane based on a Farman design, and the Factory accepted that also for production. This machine, called the Farman Experimental, was also redesigned and rebuilt to emerge as the FE2, with a gun mounted to fire over the rounded nose of the nacelle; in other words, it was designed in 1913 to be a standard fighter aircraft for the Royal Flying Corps. But it was the reconnaissance task that the Corps was set up to carry out and in 1913, and even in late summer 1914 nobody in the Military Wing was thinking in terms of aerial fighting between aeroplanes. Somebody else was.[16]

The fourth component of the great scheme was a Central Flying School. With this was coupled a plan for a reserve. The School was set up at Upavon. The Army was circulated, through the usual channels, with the news that any officer or NCO who could produce the Royal Aero Club Certificate of Proficiency in flying could claim to have the cost repaid up to a maximum of £60; surprisingly, this immediately became the standard price charged by reputable commercial schools for a flying course. The standard test included taking off, climbing to a height of a thousand feet, flying in a figure of eight, and then landing. Solo of course: nobody had yet invented dual control.

Any officer or NCO who thus qualified could apply to join the Royal Flying Corps. Those selected would be posted to a course at the Central Flying School. This was not initial flying training. It was assumed that any man who arrived at Upavon could already fly, at least up to the Royal Aero Club standards. This was to be a school of military flying, a matter of application of principles. The aim was to turn a man who could only get the machine off the ground, around a set course and back down again, into a military pilot who could select a field, land, dismount, run smartly up to the general, salute and tell the great man exactly where the Uhlans were, in what strength, whether they were accompanied by guns or encumbered by wagons, whether the horses were fresh and any other details about the enemy or the terrain which the prudent commander would want to know.[17]

It was for this course, and for attachment to the Royal Flying Corps, Military Wing, that Trenchard volunteered in late 1912. The demand, as we have seen, was for men who were experienced soldiers and qualified by rank and experience to command small detached units. Trenchard was elderly for flying: at 40 he was only just inside the age limit. But he qualified by the standards of the time, and was posted to Upavon, who probably thought they were lucky to get him and made him Assistant Commandant on the spot. There has always been an element of doubt, cultivated by Boyle, about whether Trenchard could in any sense be considered a military pilot. But that is quite irrelevant. What he could do, and what they wanted him for, was to run a unit. He had done that in South Africa, and in Nigeria, and lastly for about two years back with his regiment in Northern Ireland.

A much more typical student at Upavon was Lieutenant Dowding of the Royal Artillery. Dowding read the call for volunteers in 1913 when he was serving at a coastal battery in the Isle of Wight, a less than thrilling occupation. He got his R. Aero. C. certificate, was called to Upavon, went through the course and was pronounced satisfactory. He saw his name placed on the list of the Reserve of the Royal Flying Corps, and went back to his battery, ready to leave it and go to a flying post whenever he might be called forward, presumably to replace casualties during a war.[18]

Trenchard found other staff already in post at Upavon. Their Lordships of the Admiralty had looked at the proposed Royal Flying Corps plan, and allowed that it was very interesting. They were getting along with aviation on their own account, having been prompted from outside. A benefactor had offered free flying instruction for four naval officers in 1911, and the offer had been accepted, at least for three naval officers and one Royal Marine. The Navy had now started its own flying school, and kept one of the original naval officers to run it, a man called Samson.

However, if the Army wanted co-operation in setting up a flying school the Navy were willing to supply one captain, and two others of their original four pilots. One was the Royal Marine Major, Gerrard: the other was Lieutenant Longmore, shown in the early photographs wearing two and a half rings, because that was the badge of a lieutenant of more than seven years' seniority: the hybrid rank-title of Lieutenant-Commander had not yet been invented. The Navy also sent a young Assistant Paymaster, named Lidderdale, a sub-Lieutenant as we would say today, to keep their books in order, and an engineer officer, Engineer-Lieutenant Randall, to be Inspector of Engines. They were also willing to offer up a genuine four-ring Captain RN, G.M. Paine, as soon as he could get his RAeC Certificate.

The Army provided a remarkable man to be Quartermaster. Corporal H.H. Kirby, who already held the DCM, had won the Victoria Cross in South Africa for bringing in a wounded man under fire. He had continued as a junior NCO in the Royal Engineers, and moved up the ladder of rank in the usual way for the exceptionally able private soldier of the time. In April 1911 he reached the pinnacle of a career of this kind, and was appointed, as *Honorary* Lieutenant, to be Quartermaster to the Air Battalion at Farnborough. From this he was posted to the new Central Flying School. In the best-known photograph of the staff of the CFS he stands four square, bull-necked, uncompromising, almost terrifying; he served to 1928, retired as a wing commander, returned to duty in 1939 as a group captain, and is described by all who served with him as a perfect gentleman and a pleasure to know.

Brevet-Major Trenchard therefore set to work at the Central Flying School to turn it into an efficient Army unit, however much the Navy might wish to interfere. But Brevet-Major Sykes went on promotion to Farnborough to organize a Military Wing. There was already a unit which became 1 Squadron, and handled the airships which remained to the Army. Not for long. By the spring of 1914, the Army had agreed to hand over all the airship work of the Royal Flying Corps to the Naval Wing. More squadrons were formed at Farnborough. In February 1913, 2 Squadron took up a permanent station at Montrose, and must have flown there piecemeal and in stages, because when in August Captain Longcroft and acting Lieutenant-Colonel Sykes flew there in two stages, the first hop from Farnborough to Alnmouth set a record for a non-stop flight with a passenger. In November Captain Longcroft managed the flight non-stop from Montrose to Farnborough by way of Portsmouth.[19]

In summer 1914, Sykes had four squadrons ready, more or less, to go to war and be useful (see Table 1). He had, or thought he had, a verbal promise from Henderson that when the war started, he should go overseas in command of the Military Wing. But when the balloon went up, in August, Sykes heard, and probably not for the first time, that stirring cry of the British Army, raised always in times of stress: 'It's all been changed.'

Notes and References

1. Andrew Boyle, *Trenchard*, Collins, 1962, a biography written in close contact with the subject, is the primary source for Trenchard's life and opinions. For Boyle, Sykes is a villain.
2. F. Guggisberg, *The Shop*, Cassell, 1900, pp. 9, 14, 87, on admission to Woolwich. A. Forbes, *A History of the Army Ordnance Services*, Medici Society, 1929, on the development of the Corps. Note that in reading or writing of the period before 1939, it is not sufficient to say that a man is an Army officer, without specifying whether he is a Regimental officer or in one of the Corps, since the social origins and training were so very different.
3. See H. Mayhew, *Morning Chronicle*, 9 November 1849, reprinted in Thompson and Yeo (eds), *The Unknown Mayhew*, Penguin Books, 1973, p. 151.
4. B.W. Kelly, *The Conqueror of Culloden*, R. & T. Washbourne, 1903, pp. 121 seq.
5. Correlli Barnett, *Britain and her Army*, Allen Lane, The Penguin Press, 1970, p. 308.

6. For an account of this war, see Thomas Pakenham, *The Boer War*, Weidenfeld and Nicholson, 1979. Here, Buller is the hero and Kitchener and Milner the villains. For a British War Ministry bringing out its old plans for a second run, see p. 19. Henderson's contribution is summed up on p. 540. The visibility revolution dominates throughout.

7. U.S. Grant, *Personal Memoirs*, Sampson Lowe, 1885, p. 190.

8. Frederick Sykes, *From Many Angles*, Harrap, 1942, is the main source for Sykes' life. Sykes is his own hero: no villain is identifiable.

9. Walter Gorlitz, *The German General Staff* (trs. Battershaw), Hollis and Carter, 1953, *passim*. M. Fagyas, *The Devil's Lieutenant*, Anthony Blond, 1970, gives a readable picture of the career stresses of Continental staff selection and training.

10. R.F. Weigeley. *History of the U.S. Army*, Batsford, 1965, has illustrations of US Negro troops about 1880, wearing the Prussian helmet. British police still wear it.

11. e.g. H. Spencer Wilkinson, *The Brain of an Army*, Macmillan, 1890.

12. See Brian Bond, *The Victorian Army and the Staff College*, Eyre Methuen, 1972.

13. The standard life of Lord Esher is: Peter Fraser, *Lord Esher, a political biography*, Hart Davies and MacGibbon, 1973. The first chapter provides a convenient summary.

14. P.W.L. Broke-Smith, *The History of Early British Military Aeronautics, passim*, Library Association Academic Reprint, no date.

15. Broke-Smith, *op cit*, p. 27. For the story as the Wrights saw it, see Fred C. Kelly, *Miracle at Kitty Hawk*, Farrar Strauss and Young, New York, 1951.

16. See *Brassey's Naval Annual*, 1913, pp. 161 et seq.

17. See John R.W. Taylor, *C.F.S., Birthplace of Air Power*, Putnam, 1958. Compare Arthur Longmore, *From Sea to Sky, 1910–1945*, Geoffrey Bles, 1946, esp. Chapter 1.

18. Basil Collier, *Leader of the Few*, Jarrolds, 1957.

19. The official *Short History of the Royal Air Force* (otherwise referred to as AP125), 1936 reprint p. 13, gives the ranks as here. This single volume is much easier to handle than the six volumes of Raleigh and Jones, *War in the Air*, of which it is an abridgement.

Chapter 3

The First War

The Opening Days
The First World War started when the Austrians declared war on Serbia on 28 July 1914. France and Germany both began mobilization on 1 August. Germany declared war on Russia on that day and on France on 3 August. Britain declared war on Germany on 4 August.

The general opinion of the French staff was that the Germans would repeat their attack of 1870, and come south of the impassable Ardennes. Therefore their main body was on that frontier, and put in a spoiling attack, which didn't go very well. There was also a current of thought in France which anticipated a German attack north of the Ardennes, so a small part of their army was opposite the Belgian border. The Germans read the French mind and decided to come in well to the north of the Ardennes: this was known as soon as the mobilization began and the trains started to roll. At this point the Kaiser lost his bottle and wanted to try again for peace: his armies had not yet started shooting. His generals pointed out that the railways were already committed and to stop now in the middle would leave such a tangle that not even the General Staff could unravel it and set the German economy going again. A relatively small German force was left on the eastern frontier to contain the Russians.

The Germans knew that coming in through Belgium would bring in the British. That was irrelevant. The British Army was small and not very effective – even the Boers had beaten them.

Their mobilization would be slow. But by 21 August the Germans in Belgium began to see aeroplanes overhead with the Union Jack painted on their undersurfaces: Sykes had not yet invented the roundel. On August 22 the German vanguard reported that they were in contact with strange troops, men dressed in an unmilitary uniform of a dirty mustard colour, baggy and well pocketed, with comfortable flat hats, who dug in whenever they stopped, erected barbed-wire fences in front of their positions, opened rifle fire accurately at unheard-of ranges, and had cavalry who marched on foot and only got on their horses when in contact with the enemy. They behaved, the Germans complained, exactly like Boers and this was thought to be unfair. The British had arrived, early.[1]

When Wilson had arrived at the War Office in 1910, he found that the mobilization plan anticipated that the first British troops, in an emergency, might begin to arrive on the continent at the end of a month. By the time Wilson had finished, the time scale called for the six infantry divisions to be in France within a week, the cavalry division the day after, and the artillery the day after that. He had been very pleased with his plans, and explained it all to his best friend, a French general named Foch. On one of his regular cycling holidays on the French frontier, Wilson, the veteran of so many Staff Rides, laid his mobilization map with his planned movements marked, at the foot of the French war memorial at the battlefield of Mars le Tour. We are not told who picked it up, but it seems not the Germans.[2]

The British never learnt the difference between the General Staff, the War Ministry and the Command of an Army. Therefore as soon as the British Expeditionary Force moved, the heads of the General Staff left their desks and went off to be the field army staff. The Chief of the Imperial General Staff, Field-Marshal Sir John French, became Commander-in-Chief. His Chief of Staff was Lieutenant-General Sir Archibald Murray, who had a tendency to faint when under stress. His deputy was Wilson, who had been a prominent mover in all the staff talks since 1906 and was virtually responsible for what was happening now. The case of Colonel North has accustomed us today to the way in which comparatively junior officers can find themselves in positions of power and exercise international influence well beyond what their ranks would appear to command. The Quartermaster-General was Major-General Robertson.[3]

The commanders of the two Infantry Corps were nominated well ahead. One of them was Lieutenant-General Grierson, who inconsiderately died on the way to the Channel and was replaced by Smith-Dorrien, with whom French, to put it mildly, did not get on. The other was Lieutenant-General Haig: French owed Haig two thousand pounds, and knew that Haig, who had married a lady-in-waiting to Queen Alexandra, was having a private correspondence with his friends, King George V (who exercised power of veto over General Officer appointments) and Lord Esher. French also knew that Haig, Wilson and Robertson each wanted his job, and that the two latter at least were cleverer than he was. The amount of naked ambition, self seeking, malice and intrigue within that headquarters makes the quarrels between Montgomery, Eisenhower and Patton look like tiffs over the tea cups at a Mothers' Union meeting.

The British were in action at Mons on 22 August, and then retired along with the rest of the French. Sir John French disliked his allies as much as he did his subordinates, and felt highly insecure. He had four infantry divisions and the one cavalry division in the field, the other two infantry divisions he had brought over being left to guard the Channel ports and his lines of communications. By 4 September, he decided to take his army out of the line, to some such convenient base as Brest, and fight his own war. He told London. As a result the Secretary of State for War came out to Paris. The Minister, appointed since the outbreak of war, was Field-Marshal Lord Kitchener, and he arrived in uniform, which French thought gave him an unfair advantage. What exactly he said to French is not known, but there was no further talk of pulling out.

The great father of the German General Staff idea, Clausewitz, said that if an army didn't have a war for a generation, then officers ought to be sent to make guest appearances in other people's wars and learn what it was all about.[4] The Germans had not done this since 1870. The General Staff officers passed their service entirely in seclusion, only spending a short attachment to their original regiments at each step in rank. The German High Command lost touch with their subordinate Army headquarters, and finally sent out a General Staff Officer, Colonel Hentsch, to tour them in a car with a wireless set (both new tools for the soldier) and find out what was going on. Hentsch was horrified at the appearance of infantry who had been marching for three weeks, and decided that they could do no more. (The British and French commanders, hardened by wars in South Africa and Indo-China, knew better.) The German armies, now descending on Paris from the north, must abandon the attempt to take the capital, and pull back east to re-form and concentrate.

The High Command took the advice, and ordered the armies to change the direction of the marches, all so carefully planned years earlier. The order went out during 4 September, and the British and French dawn air sorties reported the next day that whatever the forward patrols might be doing, the main bodies of the German Army were changing their direction of march. What happened then was just what the General Staff had worked so hard to avoid: the columns collided, mingled, got into a tangle. The two right flank German armies got separated. The French, carrying the British with them, counter-attacked at what is known as the Battle of the Marne. The position was stabilized, and the serious business of intrigue at BEF Headquarters could begin.

The Military Wing
It is likely that the spotting of the German change of direction by the airmen up on the morning of 4 September was not merely a decisive incident but also that it was the only time during the war that the Air Arm actually affected the conduct of major affairs. Be that as it may, it is worth looking at how the aeroplanes got there, and who took them.

Sykes had organized the Military Wing, and thought he would command it. At the last moment, it was decided at the War Office, it is not clear by whom, that the Air Arm needed someone higher than a Lieutenant-Colonel to represent it at BEF Headquarters. A Brigadier-General was probably the lowest level who could hope to have any notice taken of him. The pattern of other elements of the General Staff was followed, and the Director of Military Aeronautics went into the field as a commander.

Whitehall in that first week of the war was full of young gentlemen looking for war jobs which would take them with the Army to France or, sometimes, honourably keep them out of France. One such young gentleman came into the War Office to see his friend David Henderson, and ask for a job. A rather elderly young gentleman at forty, the Honourable Maurice Baring was the fourth son of Lord Revelstoke. He had been at Eton and at Trinity College, Cambridge, he had spent a time, as was the custom of his class, in the Foreign Office rather than in the Army, and then had become a journalist, a war correspondent for the *Morning Post* at the siege of Port Arthur, and a political reporter in Russia. He had published two dozen books, travel writing, translations, novels, even poetry. He was by no means a nonentity, but was fairly well known in establishment and literary circles, and had some influence. And he had seen war.[5]

Henderson thought he could oblige. He told Baring to sign here, and that he was now an officer in the Intelligence Corps. For in those days a general's word was law. Baring's French was good and he could come with Headquarters RFC as interpreter. Next day the Headquarters staff set off, Lieutenant-Colonel Sykes as Chief Staff Officer, Major Brooke-Popham and Lieutenant Barrington-Kennet as administrators, and Captain Geoffrey Salmond as a General Staff Officer Grade 2, that is in Air Operations (his brother John was going out as commander of 3 Squadron). They stayed at Folkestone that night and Baring tried on his uniform. He could not put on his puttees, and none of the others was able to do it for him till the general obliged. There was some discussion as to whether Baring ought to be a second lieutenant or a lieutenant. Eventually the Staff decided in their corporate wisdom that he ought to be a lieutenant, and someone found the pips to sew on his sleeves. So they went off to war as a friendly informal family. From his diaries and letters, Baring wrote after the war his account of the proceedings.

Sykes had concentrated the four squadrons at Dover, and they flew over on 12 August, meeting the Headquarters at Amiens. From now on for some time during the retreat, until the line stabilized, there is some confusion between the Headquarters RFC and the aerodrome where the aeroplanes were, or at least one of the aerodromes. Mobilization was effective; officers on the RFC reserve who were not actually going to France with their units, like Lieutenant Dowding, were called to join the squadrons. Lieutenant-Colonel Trenchard was left at Farnborough to tidy up, and to organize the flow of replacements for casualties and of new squadrons. This he saw as a non-job for which he had been left inadequate resources. In reading accounts of this first phase one must bear in mind that everybody thought that they would be home by Christmas. This was not just a journalistic phrase. It was good General Staff doctrine. The war would entail only the movement to the first decisive battle, which would finish it all off. The only man who seemed to understand the future was Kitchener: he remembered that he had needed half a million men altogether to settle the Boers even if he never had more than half that in the field at any one date, and there were a lot more Germans than there had been Boers. A big deal is often made of Kitchener on one occasion calling for fifty more squadrons of aeroplanes, but at that time he was calling for apparently impossible amounts of everything.

The Flying Corps made contact with the Germans before anyone else. The first British soldier to be wounded was Sergeant-Major Jillings of the Grenadier Guards, flying as an observer, who was shot through the seat of his aeroplane. This wound, we are informed, had no influence on the formation of the Air Ministry.[6] Mr Jillings recovered, and retired as a group-captain: his son also achieved senior rank.

We are now deceived by such films as *Those Magnificent Men in Their Flying Machines* into thinking of early flying as something pleasant and without tears. But at the beginning of the war flying the usual military aeroplane may best be compared with riding a rather dilapidated motorcycle with balding and wrongly inflated tyres over a trials course in freezing fog and a gale force side wind, refreshing yourself with draughts of neat castor oil. Just getting it off the ground and flying it in the direction needed was difficult, and being shot at can have added only marginally to the anxiety. One problem was weight.

Our folklore picture of the time is one of pilots taking to the air armed with rifles and revolvers. To understand the facts of armament, it is worth looking at a photograph of the standard reconnaissance machine, the BE2, and asking where in the absence of synchronizing gear you would use even a rifle, or stow it. A rifle weighed 10 lb, a pistol 2 or 3 lb. The only machine gun available in 1914 was the Vickers, the British form of the German Maxim. Here the gun weighs 40 lb, without the coolant: the ground tripod also weighs 40 lb, so an aircraft mounting would come to about the same. With the ammunition, this armament adds up to the weight of a small man. To get enough performance to do the job, crews were leaving behind the rifles and revolvers they were issued with, and the armour plating fitted around their seats: Mr Jillings was wounded as he was simply because he and his pilot had deliberately weighed the odds and committed an act of calm cool courage, in dumping all protection.

The first fighter aeroplanes of a kind were the Maurice Farman Shorthorns of 4 Squadron, which were sent out at the end of September, pusher aeroplanes with a machine gun mounted in the front of the nacelle. The RFC didn't find them much use, simply because they hadn't got any performance to spare. The Lewis gun had been tested in 1913, and recommended for the RFC, but the first Army buy didn't start to arrive till into 1915. Back at Farnborough, the FE2 had been originally designed to carry a machine gun, but there was no quantity production under way yet. After this opening phase, the development of aircraft for the RFC followed the state of the art. What was needed in terms of power, manoeuvrability and armament was easy enough to see, but, like the Kite balloon, it wasn't always easy to build it. It was, for example, easy enough to see that the flexible front gunner in the pusher hadn't got much future: the clever thing was to invent a synchronizing gear to save the weight of the extra man and let the pilot aim the gun by aiming the aeroplane. Mynheer Fokker got there before Mr Constantinescu: just.

The Naval Wing
Other people than the Military Wing had been thinking for some time about other ways of waging war from the air than reconnaissance and spotting for artillery. In an army where you might find a colonel clinging to a kite several hundred feet above Salisbury Plain, while being pulled along by a team of galloping horses, it is not unexpected that other officers had similar brainwaves.

During the first decade of the century, a Royal Engineers officer named Swinton began to write and to publish what we would now call science fiction short stories with a military application; his pen name was Ole Luke Oie. One of these tales, *The Joint in the Harness*, went into Blackwood's in January 1907: that is, it was written in autumn 1906. It outlines a bombing raid on a party repairing a damaged bridge. The aircraft are obviously what we would now call microlites, that is they are based on the work of Lilienthal rather than of the Wrights. They are brought to the forward position in crates, unpacked and put together. The bomb load is 20 lb, besides 'incendiary leaves'. All the obvious points are covered, navigation, bomb aiming, the expected wastage on take off and on landing back. The aviators are young, and dressed in a kind of 'hybrid naval uniform'. And, most modern touch of all, the target is not the bridge, but the pile-driver which is on the forward side and cannot be replaced unless the bridge is, somehow, repaired without it. Casualties were five machines out of twelve, a reasonable estimate.[7] In the following year, H.G. Wells wrote *The War in the Air*, which envisages bombing raids by Zeppelins carrying aircraft, motorized hang-gliders again, which are detached to attack battleships. The ideas were in the air, and it was the Navy, rather than the Army, which took them up.

The Navy didn't scorn reconnaissance. That was the basic need; they had already, before 1914, begun to try launching aeroplanes from warships. By 1915 four *Arethusa* class cruisers of the Harwich Force were fitted with flying-off platforms to take French monoplane fighters to see off Zeppelins, which shadowed the warships in very much the same way as the Condors shadowed convoys in the Second World War. The Condors were seen off by Spitfires carried on merchant ships and catapulted into the air: the aeroplane then, just as in 1915, landed in the water, and the pilot trusted to luck that he would be picked up. In the First World War there were many other experiments in carrying seaplanes, which were then expected to land in the water, to be hoisted back aboard the carrying vessel. In the Harwich Force, the aeroplanes were not a success because they did not have the ceiling to tackle the Zeppelins.

The Navy also used observation balloons towed behind ships. This was awkward because the balloons had to be inflated on shore and towed out. And they had their own airships, some of which they had built themselves and some of which they took over from the Military Wing. The real requirement, however, was to beat the Zeppelins. By the end of the war, the sailors were in the habit of taking out four-seater flying boats with three or four gun positions and stalking Zeppelins among the clouds; they had some successes, although it seems not to have been unusual for the flying boats to have to come down in the water and taxi home, which might take days.[8] The best thing was obviously to strike at the Zeppelin bases.

There was then an interlude of some importance. The Navy, when they were told about it, which was quite late in the day, didn't think much of the Army's going to France: it knocked out their idea, which was that when the expected war came the Navy should pick up the Army and put it down in the Baltic, on the Prussian coast where there were plenty of wide flat beaches. How this was to be done in the face of the High Seas Fleet, and how the Army was then to be supplied, they did not explain. As their Lordships couldn't have the Army for their foray, they decided to make a shorter excursion and put down what troops they had, the Royal Marines, at Antwerp. They eventually had to pull out, leaving a large number of their private naval army to be interned in Holland.

But while they held Antwerp, the Royal Naval Air Service, as the Naval Wing RFC had become on 1 July 1914, sent two squadrons there. These carried out bombing raids into Germany. The first, on 22 September, was to attack the Zeppelin sheds at Düsseldorf. A second raid, on 8 October, hit the railway station at Cologne and destroyed the new Zeppelin Z1X at Düsseldorf, together with – echoes of Swinton and the pile-driver – the airship station's machine shop and erecting sheds. By this time the Germans were closing in on the airfield at Antwerp, and the squadrons were pulled out.

Meanwhile the RNAS had established itself at Ostend and then at Dunkirk to try to deny the Belgian coast to the Zeppelins and to the German Navy. The commander of this force was Commander Samson. It had aeroplanes, and was also given a handful of armoured cars to recover crashed aeroplanes and to secure aerodromes. The Boer War, with open mobile fighting, had also obsessed the admirals. Samson carried out a piratical campaign all of his own till the line hardened, the forerunner of Popski's and a multitude of other private armies in the Second World War.[9]

Among the many well-wishers of naval aviation who had got in on the act and had been given reserve commissions, was the proprietor of Supermarine Aircraft, Pemberton Billing. Billing had sold his company on joining the Navy. Now he organized a remarkable expedition, in which, after arrangements with the French, four aeroplanes were taken down to a base at Belfort. From here, they launched a raid on the Zeppelin factory at Friedrichshafen. On 21 November three of the four aeroplanes managed to take off, each carrying four 20-pounder bombs. They flew across Lake Constance, playing ducks and drakes with Swiss neutral airspace, and found the target. They hit the gasworks and set it on fire, they damaged the other construction sheds and a Zeppelin under construction. One raider came down in French territory on the way back; the other two aeroplanes got home to Belfort, where Lieutenant Billing had them packed up and taken back to England.

Royal Flying Corps, France
During the movement of the BEF to the northern part of the allied front, the two Corps were elevated to the status of Armies. Each was given one Wing of two Squadrons of aeroplanes for its own. The most senior RFC officer in England, Trenchard, was brought out in November 1914 to command one Wing, with 1st Army under Haig. The other Wing was put under John Salmond.

In December 1914 Brigadier-General Henderson, after much pleading, was given command of an infantry division, a major-general post. The aim of every officer was command of his regiment, or at least a battalion of it, in the field. If that could not be arranged, a division was next best thing. About this time, Major-General Capper was brought back from the Staff College at Quetta to command a division, and was killed leading his men in the assault. It may have been a splendid military end, but it does seem a shocking waste of talent and training. Henderson went to 1st Division, and Baring went with him. What struck Baring most forcibly was that the Flying Corps rode in cars, but the infantry walked. Everywhere. There is no doubt that Baring would cheerfully have walked after Henderson into enemy fire, but walking from the office to lunch was a serious matter. He did not suffer for long.

In a few weeks BEF prevailed over the War Office and over Henderson's military ambitions, and he was brought back to BEF Headquarters. He was, however, allowed to keep his new rank of major-general. While Henderson was away, Sykes had acted as Commander RFC. Now Henderson was back, he reverted to Chief of Staff. There had been other changes. By now the staff at BEF Headquarters was in a state of some disarray. There was a gulf between the Wilson men and the rest. Sykes was a Wilson man. The first man to gain from all this was Robertson, who in February 1915 found himself promoted to lieutenant-general and appointed Chief of the Imperial General Staff. This he may have considered as being kicked upstairs. He was soon a full general, but he was now removed from the scene of action. Wilson followed, and became the BEF representative at the French General Staff. Sykes thus found himself isolated. In April 1915 Henderson was finally recalled to London to become again Director of Military Aeronautics. When the dust of conflict had died down, Trenchard was a Major-General and Chief, Royal Flying Corps, at Headquarters BEF. Sykes, plainly, was out.

But what to do with Sykes? He was too valuable and well known a man to be simply sacked. So the Army did what the song said, they sent him to the Dardanelles. This theatre of war was just getting active, and the air warfare scene was in some confusion. The original naval commander, Admiral Carden, who was committed to forcing a way through the Narrows without land support, had decided that he would need aeroplanes for reconnaissance and spotting for his battleships carrying out massive shoots in waters fairly thick with submarines. He asked the Admiralty for air support, and to be fair, the Admiralty took him seriously and sent out their best man. They sent Samson. There were two problems. They didn't send out any adequate number of serviceable aeroplanes. And while they were all on the way, Carden died and was replaced by Admiral van Roebeck, who didn't want aeroplanes and couldn't see what use they would be.

This was the time of the great Army assault on Gallipoli. Samson got his aeroplanes flying, after a fashion, which was a task and a half in itself, and went around touting for trade. The Army were very glad to see him. They were used to aeroplanes, and knew exactly what they wanted them to do. So Commander Samson went to work spotting for the Army from an airfield on Imbros.

Lieutenant-Colonel Sykes went to the Dardanelles, saw the confusion, and wrote a report. Their Lordships approved of it, and Sykes, naturally, was sent out to the Dardanelles to implement it. For the occasion, he found himself an acting colonel in the Royal Flying Corps, an acting colonel in the Royal Marines, and a temporary and local wing-captain in the Royal Naval Air Service. Sykes is convinced that he did a great deal of useful work in reorganizing aviation in the eastern Mediterranean. Samson – and this is almost the only occasion where any one of the named actors in this story says anything very much about another – says that his arrival merely meant another carbon every time they typed a report. The truth is probably between the two accounts.[10]

The Army evaluated Gallipoli. Acting Colonel RFC, Acting Colonel RM, Acting Captain RN Frederick Sykes was brought home, and on 1 February 1916 he reverted to his substantive rank and was posted, as a captain in the 15th Hussars, to the headquarters of a Territorial Cyclist Division at Colchester. At this point he must have reflected that Trenchard, or Haig, or somebody, had a long, long arm.

The Course of Events
Things changed at General Headquarters, BEF, during 1916. Robertson was in London as CIGS, and now Wilson was exposed, as Sykes had been. He had been removed from the BEF where he might have affected operations, and sent to be liaison officer with the French *Quartier Général*. During the next eighteen months he was moved, first to command a Corps in France, which brought promotion to lieutenant-general, and then to England as GOC Eastern Command, a backwater in his terms. At the end of 1916, Field-Marshal French had a heart attack, and the way was clear to retire him. Haig was promoted in his place.

There is no doubt that Trenchard and Haig were on good terms, to say the least, yet it is strange that very few of the great men of the time, including Haig, have very much to say about Trenchard in their memoirs. He is almost always mentioned with some such phrase as 'General Trenchard, who is head of the Flying Corps', as if he were hardly a household name, and certainly not a dominating figure.[11] Only General Charteris, Haig's Director of Military Intelligence, remarks in his published diaries that 'if Douglas Haig were to go, I think there is only one man with the strength to succeed him, and he is Trenchard.'[12] But Charteris did not publish this till 1931, and there is other evidence that his words are not entirely reliable: besides, in 1916 Trenchard was merely another major-general, not Staff-trained and hardly a contender for such a highly political post. He was a major-general only because he had taken over command from Henderson, and Henderson held that post in the rank of major-general for the last few months because he had been promoted to command his division.

As far as any picture comes through of Trenchard, it is of a patient plodding reliable work-horse, running his own little bit of the BEF, just like the lieutenant-generals who were Chief Royal Artillery, Engineer-in-Chief, and so on. Trenchard was junior in rank to these officers, but functionally their equivalent. He was only a major-general, and surely that was well beyond anything he had expected in earlier years, but he was an important major-general. He was not of the same level as the Chief of Staff or the Director of Intelligence or the Quartermaster-General, but even if he did not have direct access to Haig, Haig had direct access to him, without going through any intermediate generals.

Any novelist knows, and many human scientists know, that if we want to describe the personality of a man, and explain how he got the way he is, the most important thing to take into account is his job, what he does all day and every day. Not all biographers know this, and autobiographers usually seem to make some effort to brush it over. Nevertheless we may as well ask, what was it Trenchard did all day. We are lucky, because Baring goes to some pains to tell us just that.

Baring had one great advantage. Because of the confusion of the earliest days, when the one open field held around its margins the squadrons and the Corps Headquarters, with its workshops and its offices and its messes, the RFC Headquarters was in a different place from the rest of the General Headquarters, BEF. Its mess was always a separate household, a different family, with Trenchard at the head of the table. It was a busy, cheerful place and there were always visitors, guests, middleweights at least. There were soldiers, politicians, artists, writers, to be shown how the RFC worked, what its commander was like, how valuable its contribution was to the war effort. These guests went away highly impressed, and spread the word. What Trenchard may well not have realized was that however pleased with the RFC picture these visitors were when they went away, they had in the first place not come there to see the RFC, or its Commander: they had come, the younger politicians and the writers and the artists and the sons of the Establishment families, to see their friend Maurice Baring.

Trenchard's job was very like that of his functional equivalents, the Chiefs of Arms. He went to meetings at BEF General Headquarters: staff meetings, planning meetings, progress meetings. According to Duff Cooper, Trenchard visited General Headquarters once a week to report on the Air War, but he does not say whether this was a regular full-scale presentation, or whether

Trenchard merely attended the C-in-C's weekly meeting, where he appeared on the agenda as contributing after the more senior generals had spoken.[13] Sometimes the meetings were in London, meetings with Director-General of Military Aeronautics, or the President of the Air Board. More frequent were meetings at his own Headquarters, with his internal staff or with his brigade commanders: the wings at the disposal of each of the five armies were now brigades, and their commanders were now brigadier-generals.

When he was not at meetings, Trenchard did what the other Heads of Arms did, and spent a great deal of his time visiting the subordinate formations: but with a difference. His colleagues kept visiting at a high level, Army, and perhaps Corps. Trenchard visited squadrons. The Chief, Royal Artillery, for example, did not talk with the troops: he might talk *to* them, in set piece speeches at parades and great occasions. He might occasionally be affable to such low life as battery commanders, but like other generals he did not associate with the men who did the actual fighting. This lack of contact may well have been, formally or informally, policy: the Army was haunted by the memory of Buller who had in Natal not once but three times been on the brink of shattering the Boer army before him, and each time had called off the final assault because he could not bear the sight of his own wounded being brought back past him.

But the RFC was different. The men who did the fighting were few in numbers, and almost all were officers. Trenchard visited squadrons, and let it be known early in his command that when he did this he did not want brigade staff getting in his way. The pilots got used to him, arriving by road or sometimes by air in days when flying even as a passenger was a dangerous adventure. He would move around the flights with the minimum of due ceremony, and eat, sometimes sleep, in their mess. He didn't say much, except for the occasional not very agreeable grunt which gave him his nickname of 'Boom'. But he listened to them, and the aftermath would prove it. So they talked to him, or in his presence, at length and with little pretence. He would come, quietly, when something was going to happen, and he would come, just as quietly, when the something had happened and the squadron had had a bad, or even a good time. The pilots felt that they knew him. He was their commander, they were his pilots, in a way in which no one else above battalion commander had his own men. It was Trenchard who created the atmosphere of the Flying Corps, and bequeathed such slogans as 'Shut the hangar door' or 'Close up at table'.

And always behind Trenchard was the figure of Baring with his notebook. Trenchard had inherited Baring from Henderson, and after trying him out decided he was worth keeping on. On these visits, Trenchard asked questions, and Baring wrote down the answers. Then back at Headquarters, Trenchard and Baring would work over the notes and Trenchard would turn them into the form of questions and instructions for the Staff. There was a shortage of this lubricant, the transport wasn't available from Corps, the method of signalling to batteries could be improved. And in three days, Baring would have collected and collated the answers for Trenchard to see, and if the answers weren't acceptable, then Trenchard would want to know why. Trenchard was a high level progress chaser, and the results were known to the pilots. Trenchard was the man who knew everything that was going on, and if the pilots, obsessed as they were with the performance of their machines, had a real complaint, then Trenchard would see to it. It was in these days that the legend of Trenchard as the Father of the Air Force was hammered out.

Home Affairs
The pilots had a real complaint about their machines. They were not, on the whole, very good, either as aeroplanes or as fighting machines, and it showed. Lieutenant Pemberton Billing came home from the attack on Friedrichshafen, which was incidentally carried out with Avro 504s, in a state of some passion. He obtained his release from the Navy, and went into Parliament, being elected on a 'better aeroplanes' ticket. There he began to campaign for just that. The months after the outbreaks of war had been a happy hunting season for aeroplane makers, and for engine makers. Buying agents for both Army and Navy had toured Britain, looking for firms who could make aeroplanes or engines, and handing out contracts. The Army were looking mainly for subcontractors to make parts for the FE2 and the BE2, as well as for engine components. The Navy were also looking for men who could design or had designed new aeroplanes for their new tasks. The Navy were already employing such designers as T.M. Sopwith or A.V. Roe. The

two ministries were competing for scarce resources and bidding against each other.

It must be remembered that while maintaining an air force had proved to be massively expensive, much of the expense was self-generated. At the beginning, all the generals wanted was a few machines to carry out reconnaissance while the armies were mobile, and then to do artillery spotting when balloons were unusable in bad weather in static warfare. The next stage was to produce armed aircraft to attack the enemy's reconnaissance machines. The third stage was to produce more armed fighters to keep the enemy's fighters from shooting down your own machines. The final stage was to produce large masses of fighters simply to fight the other side's fighters in massive patrols and sweeps. Air War was being carried on not to carry out the job the armies wanted but simply to fight an air war for its own sake, or in polite terms to gain control of the air.

While bombing was conspicuous in the prewar literature, it was at first a relatively minor feature of operations. The Naval Air Service was, as we have seen, the earliest practitioner, attacking first Friedrichshafen and then the Rhine towns, as long as they had the range. A raid on Cuxhaven on Christmas Day of 1914 by seaplanes carried into range on ships and coming down in the water afterwards, had the effect of making the Germans abandon Cuxhaven as an anchorage. The Navy began to build up in Belgium and the north-east corner of France a bomber force of its own which carried on a continuous attack on the Zeppelin sheds along the coast. In the usual manner of any undertaking directed by Winston Churchill this force mushroomed, until when he left office the Admiralty handed over to the War Office fifteen armoured car squadrons, three armoured trains, several kite balloon squadrons and an anti-aircraft force of 1,500 men; all the legacy of Commander Samson.

As the Navy was responsible, as tradition demanded, for the defence of this island, it developed fighter squadrons stationed at Dunkirk and along the south-east coast. As a result the Navy and the Army were competing for air resources and even for targets with each other. At the end of 1915 the Army, that is the RFC, claimed responsibility for the aerial defence of London and positioned twenty additional BE2cs at ten stations around London. But the Naval Squadrons remained where they had been, and there was still confusion as to who raiding Zeppelins belonged to. In German thinking, the Zeppelins belonged to their Navy, since the Army tended to use the Schutte-Lanz: but to British civilians they were all Zeppelins.

The purchasing chaos continued into 1916. In the spring of 1916, a Committee of Inquiry under Mr Justice Bailhache examined the administration of the Royal Flying Corps. In February 1916, a permanent sub-committee of the War Cabinet had been formed to coordinate air supplies for both services. It became clear that the two services held different opinions, especially on the question of long-distance bombing: only the Navy had been doing this. In May, this sub-committee turned into an Air Board to advise, only, on matters of policy and in particular on combined operations of the two services: this obviously refers to the Dunkirk area, where the Naval Air Service was working for the Army, as well as to the Dardanelles to which Sykes was sent. This, the first Air Board, is usually known by the name of its chairman, Lord Curzon.[14]

In December a second Air Board, usually called by the name of its president, Lord Cowdray, was set up with some executive powers. In accordance with the recommendations of the Bailhache Committee, this Board was to be responsible for the design and allocation of aircraft, while the Ministry of Munitions was to be responsible for production. The President of the Air Board was to be, legally, a Minister, and his organization a Ministry. But Lloyd George, having been first Minister of Munitions, and then Secretary for War in succession to Lord Kitchener, had become Prime Minister on December 7 1916.

Now two extra factors entered the development. One was the German Air Arms, Army and Navy. There had been a few air raids on Britain in 1915, mostly by Zeppelins at night. During 1916 the Zeppelin raids began to fade out, and aeroplane raids started. The usual target was Dover, but on 28 November a solitary naval aeroplane reached London and bombed Victoria Station.[15] During 1917 the Germans began to use the Gotha twin-engined bombers, with a bomb load of 350 kg and enough range to reach London. The first raid on Folkestone was on 25 May, and the first on London on 13 June. There were in all seven big daylight raids, and the campaign was abandoned in August, but nobody in Britain knew that it was over, at least for the time being. The press made the most of the sensation, and there was pressure on the Prime Minister to do something.

The other factor was General Jan Christian Smuts. After being a guerrilla leader in 1901 to 1902, he had gone into politics in the new self-governing dominion of South Africa. He had led a South African force in the destruction of the German garrison in East Africa. In 1917 Lloyd George called a meeting of an Imperial War Conference of representatives from the Dominions, and when that was over and everyone else went home, the South African representative, Smuts, stayed on. Lloyd George co-opted Smuts into the War Cabinet, with no political base: when urged to take a seat in the Commons, which could have been arranged, Smuts refused. He was one of the many political balls Lloyd George was juggling with, and the air raids problem was another. Other difficult balls included the two brothers Harmsworth, better known as the press barons Lords Rothermere and Northcliffe.

At the beginning of August 1917, Lloyd George persuaded the War Cabinet to appoint a sub-committee to consider (a) home defence against air raids, and (b) the existing organization for the study and higher direction of aerial operations. It had two full members, Lloyd George and Smuts. Smuts did the work and produced the reports Lloyd George wanted. Smuts, according to his biographer, 'leaned heavily on his adviser, Major-General Henderson, the Director-General of Military Aeronautics'. The first recommendation, produced in a fortnight, was that air defence of the capital should be brought under a unified command. This was done.

The second recommendation came at the end of August. It urged the creation of an independent Air Staff backed by an independent Air Ministry. The War Cabinet accepted this recommendation and appointed Smuts as head of an Air Organization Committee. Smuts continued to take the advice of Sir David Henderson, who had in July sent him 'a powerful memorandum' on the subject. The final result, in the creation of the RAF as it came out by the end of the year, is obviously the Royal Flying Corps as devised in 1910 by Henderson and Sykes, with detailed changes in the light of war experience. The War Cabinet accepted all the Committee's recommendations. The bill to set up the new structure was drafted and became law by the end of November. Hancock, Smuts' biographer, thinks that it was remarkable how quickly Smuts mastered the procedures of Whitehall negotiations: but Henderson had been doing nothing else for years.

The creation of the Air Ministry enabled Lloyd George to deal with several of the balls he was trying to keep in the air. In the first place he gained the reputation of being the man who had done something to deal with the air raids. The air raids kept on. The Germans supplemented the Gothas with Giants which were awkward to build and to handle, but which had a bomb-load of 1,000 kg. The last aeroplane raid was on the night of 19 May 1918 and the last Zeppelin raid on the night of 5 August.

In the second place, he had kept Smuts busy and out of his hair for four months. Smuts was hyperactive. He must have been a dreadful trial to Lloyd George, who had made several attempts to send him elsewhere, including offering him the command of the forces in Palestine. Now Smuts was sent off to make soothing speeches in South Wales, where Lloyd George, a North Welshman, had contrived to quarrel with the Miners' Unions and spark off a work-to-rule.

Thirdly, Lloyd George was given control of a large new piece of political patronage. He at once offered the new office of Secretary of State for Air to Lord Northcliffe, who refused it.

Lastly, it gave Lloyd George the chance to remove a large slab of manpower and production from the immediate control of the generals, whom he hated, seeing their stupidity as responsible for all the carnage in France. He didn't like the admirals very much either; they had been given the fifty battleships they had wanted, and a hundred thousand men shut up in the Orkneys, but they had not destroyed the High Seas Fleet as promised, and the Island was almost starving.

The Prime Minister's main target was Haig, both because of his own opinions and because he had made friends with Wilson. He was sure that if he could get rid of Haig the war could, by some magic means, be swiftly won without any more losses. How, he didn't know. But to sack Haig was impossible. He had been built up as the symbol of British land power, and was a massive propaganda asset. To sack him, as lesser generals had been sacked, would be seen not merely as an admission of defeat, but as a proclamation that the Army had failed.

The Army did not see it that way. A few upper-class intellectuals serving in France might believe it, but the troops knew what they had done. They had fought on the Somme against a hard, brave, resourceful, cruel enemy, well armed and well posted on ground of his own choosing, when they were themselves so green that they were fit to do nothing more complex but advance in straight lines at the walk. There through three months they had beaten him, by sheer determination, from one trench line to another, till at the end, rather than face them again on the same ground in the spring, he had retreated over a scorched earth for a dozen miles to a vast triple belt of fortification, the Hindenberg Line. In the following year they had broken out of Ypres, at a time chosen by the French, and again forced the enemy back up Passchendaele, until by Christmas, for all the blood and the mud and the rain, a man could stand upright on the ridge and see far into Belgium where the Germans awaited their assault. And in all this it was Haig who stood foursquare as a symbol of their valour and dedication. In the following year, when the Germans brought back their troops out of Russia and for the first time since 1914 launched a series of major attacks, Haig expressed the spirit of the men exactly in his 'backs to the wall' message. The real mottoes of the County Regiments, not the Latin slogans on the cap badges but the rallying cries passed from mouth to mouth, expressed just that tone of stoical pessimism – 'Die Hard, Middlesex', 'Stick it, the Welch'.

Haig could not be dismissed. The troops would not have it, and there was also the King . . . But there were other ways of rendering him harmless. During the last two months of 1917, most of his senior staff were found other posts. His intelligence chief, Charteris, says that his doctors advised him to resign his post: but Lloyd George says flatly that he ordered Haig to sack Charteris, and Haig did. The Chief of Staff, the Deputy Chief of Staff, the Quartermaster-General, the Engineer-in-Chief, the Director-General of Transport, even the Director-General of Medical Services, all went.[16] Most of them were found other posts, but such a purge, as Mr Terraine calls it, cannot possibly be shrugged off as normal rotation. Not all at once. Nobody who has not worked on a staff can understand what the atmosphere must have been like as all the Commander in Chief's immediate advisers were removed. In all this, the posting out even on apparent promotion of yet another major-general cannot have attracted any notice. Another post was found for Haig's friend Trenchard, the man his pilots loved, so that they would be sure that this was a signal mark of honour.

Trenchard was sent home to be the first Chief of the Air Staff. The new Secretary of State was Lord Rothermere, Northcliffe's brother. Lloyd George threw together these two rigid, opinionated men in the sure and certain hope that they would very soon tear each other apart. The wonder is that it took so long, and that can only be a tribute to Trenchard's patience. The pretext for the conflict was simple. Rothermere saw the Air Ministry as a temporary administrative expedient, and the Air Force as equally temporary, to be dissolved as soon as possible, and the whole matter something cleverly contrived by Lord George in order to give one or other of the Harmsworth brothers a place in the Administration and therefore political power. Trenchard on the other hand had the classical Army attitude that a war was a mere episode in a soldier's career, and as soon as this war was over he could get back to real soldiering. He had been given the task of forming a new force, and he wanted to get on with it and provide it with all its basic structures of cadet colleges and hospitals and other permanent features. Who would replace them when first Trenchard resigned in April 1918 and then Rothermere went rather than face the Parliamentary music?

The candidates were obvious, the men for whom events had created the posts. One member of the new Air Council was the Director-General of Aircraft Production at the Ministry of Munitions, Sir William Weir. Weir was head of a Glasgow firm of engineering manufacturers. He had come to notice early in the war when he took a firm line with the unions. Lloyd George had recruited him to service in the Ministry of Munitions, and there he had been placed on the first Air Board as an expert in engineering production and man management; his firm was a major supplier of castings for cylinder blocks for aero-engines. He had remained in place on the second Air Board, and could now be expected to have a wide knowledge of the problems of air defence and its supply.[17] He was given a peerage for the purpose, and as second Secretary of State for Air he appointed, as his Chief of Air Staff, Major-General Sykes.

The New Order

Where had Sykes come from? He had last been seen as a staff-captain with a territorial division. But nobody could think of leaving him there. He was very soon posted as a major to the Machine Gun Directorate at the War Office. He found himself with the usual bare office, a borrowed table and chair, a typist and a typewriter, and told to organize the new structure for the Heavy Motor Section.

Some time earlier, Ole Luke Oie, now Major-General Swinton, had taken an idea from H.G. Wells and begun a project to mount machine guns on caterpillar tractors protected with armour. At first he found little support outside the Admiralty, but at length the Army had taken it over, although Swinton made himself so unpopular by his persistance that he was sacked. The new machines, under the cover name of tanks, were now to come into service, and Major Sykes was to work out an organization for the units. This he did, and among other things, time being limited for designing an entire new uniform, he gave them a distinctive and original headgear, the black beret. Colonel Sykes was promoted and moved to be Director-General of Organization, to deal with manpower planning: we should remember that most of Lloyd George's quarrel with the generals was over the proper use of scarce manpower. Brigadier-General Sykes was given the problem of how to use women in the uniformed services, and he worked out an organization for the Womens' Auxiliary Army Corps.

But he felt that his talents were not being properly used, and so did other people. In November 1917, the Prime Ministers of Britain, France and Italy formed the Allied War Council, with a permanent secretariat at Versailles, and the military advisers were Foch, Cadorna, and Wilson. Wilson needed a deputy, and the French-speaking Major-General Sykes was appointed. He was also made that member of the secretariat responsible for Manpower Planning.

Now Sykes was to be made Chief of the Air Staff, to head the Corps he had devised seven years before, with its thousands of aeroplanes, its hundreds of thousands of men, and all its stations. But if the truth be told, Sykes had grown beyond this. He had tasted the blood of war at the top, of supreme command. He accepted the post of Chief of Air Staff, but on a part-time basis, retaining his other post at Versailles. He commuted weekly by air, when this was a risky business; once he was forced down by icing in Belgium, and several times had to be ready for ditching. Presumably he was a beanstealer impartially at each end. But he was an old, bold pilot, and he enjoyed flying. These risks were all part of the fun.

The war went on, with a part-time Chief of the Air Staff, and John Salmond called as a Major-General to command the RFC at General Headquarters BEF. In March 1918, the Germans were able to bring troops from Russia and to break out of their great defence system. The British Army put its backs to the wall, and held the Germans. The American troops were coming in force, but more important were the good aeroplane engines, the Liberty engines, they were supplying. At the beginning of August Haig, the old cavalryman from the mobile war in South Africa, with his new staff, counter-attacked, and in a hundred days of unbroken victories smashed the right wing of the German Army. In November it was all over.

Things were changing for other people. Immediately the war ended, Lloyd George called an election, the Khaki Election with suffrage now granted to women over 30. He won. Lord Weir resigned: he had only come to London for the war, and he wanted to go back to his business in Glasgow. A new Secretary of State for War and for Air was appointed, that good Liberal Winston Churchill. On 1 January 1919 he published his new appointments. Major-General Trenchard was to be Chief of the Air Staff again, and Major-General Sykes was to take a new post, of equal dignity and standing with that of the military commander. He was to be Controller-General of Civil Aviation. The war was over.

Notes and References

1. The literature on the opening days of the First World War is vast and not always helpful. Perhaps for the general reader the best is Barbara Tuchman, *The Guns of August*, Dell Publishing House, 1962.
2. Bernard Ash, *The Lost Dictator*, Cassell, 1968, is a critical rather than an adulatory life of Wilson.
3. W. Robertson, *From Private to Field General*, Constable, 1921, may be supplemented with V. Bonham Carter, *Soldier True*, Muller, 1952,

especially for its account of the War Game of 1905 which was based on the threat of a German attack through Belgium.

4. C. von Clausewitz, *On War*, is available as a Penguin Classic. See pp. 139-143 and 147 on friction in war and habituation to it.

5. Maurice Baring, *Flying Corps Headquarters, 1914-18*, has been reprinted several times by different publishers, and is available in paperback. It is simpler therefore to give references by dates rather than by pages. Trenchard's methods of work are described under August 1915. Boyle, *Trenchard*, does not describe adequately who Baring was in real life.

6. C.G. Grey, *A History of the Air Ministry*, Allen and Unwin, 1940, has many irritating defects but is invaluable. The initial engagement is described on p. 46.

7. Ole Luke Oie, *The Green Curve Omnibus*, was published by William Blackwood and Son, Edinburgh, 1909, and reprinted, either entire or in smaller collections, frequently since. The two relevant stories are *The Kite* and *The Joint in the Harness*, although the title story is an early exposition of cold-blooded operational research methods.

8. For these exploits see the Official History, AP 125.

9. C.R. Samson, *Fights and Flights*, Ernest Benn Ltd., 1930.

10. H.A. Jones, *The War in the Air*, OUP, 1928, Vol. II, p. 57, agrees with Sykes on the value of Sykes' work: Samson, *Fights and Flights*, p. 265, differs.

11. See Robert Blake (ed), *The Private Papers of Douglas Haig*, Eyre and Spottiswood, 1952. Haig edited and pruned his papers with a view to publication, and Blake has further edited and selected from the survivors; they cannot be regarded as a good source. For a few mentions of Trenchard, see pp. 216, 245, 252, 273, 280.

12. J. Charteris, *At GHQ*, Cassell, 1931, p. 273.

13. Duff Cooper, *Haig*, Faber and Faber, 1936, Vol. II, p. 212.

14. The sequence of Air Boards is frequently described, especially by Gray, op. cit.

15. The crew were Deck Officer Brandt, pilot, and Lieutenant Ilges, aircraft captain. The significance of these naval ranks is explained in the following chapter.

16. D. Lloyd George, *War Memoirs*, Odhams, 1938, pp. 1095 et seq. John Terraine, *Douglas Haig - the Educated Soldier*, Hutchinson, 1963, p. 388. The Official History, *France and Flanders 1918*, Vol. I pp. 55-6.

17. W.J. Reader, *Architect of Air Power, the life of Viscount Weir*, Collins, 1968. The title is clearly explanatory of the tone of the book.

Chapter 4

Transition

The New Beginning

It had been a long haul, from the start of the RFC in 1912, through the first Air Board in May 1916, and the second Air Board in December 1916. The Air Board's Headquarters were in the Hotel Cecil, popularly known as Bolo House: Bolo Pasha was at that period notorious as a French political swindler on a large scale. Its staff was large and regarded as a set of idle bureaucrats, which is probably unjust: as long ago as 1666, Pepys had remarked on how much care was needed by how many people just to see that the King's money was not wasted,[1] and there was enough scope for waste, and even for straightforward cheating and corruption. The choice was, as always, between a large and relatively cheap bureaucracy and a vastly overexpensive service. All this was taken over into the new Air Ministry from 1 January 1918. From 1 April it was Trenchard's Air Force. From 14 April it was Sykes' Air Force. What kind of Air Force was it?

To start with, it was a widespread empire. At the beginning, it consisted of 220 squadrons (Table 2). Nine of these are noted as 'non-operational', and may have been in the early stages of formation, since they are all given as based in the United Kingdom, and three of them are allocated roles as day or night bomber squadrons. Two other squadrons appear on the list as formed after the beginning of April, and are given as non-operational. Of the squadrons which may be assumed to be operational, the majority were in France or in the United Kingdom, which of course at this date included Ireland. Most of the squadrons were from the RFC, and are referred to here as 'Army' squadrons: the squadrons which came from the RNAS are referred to here as 'Navy' squadrons, and were all given numbers in the 200 series: e.g. 16 Squadron RNAS became 216 RAF (Table 3).

The contingent in France is the largest. The RFC contributed sixty squadrons, together with four squadrons on loan from the Navy, and under RFC control. But the Navy also had nine squadrons of its own in France, mostly of course in the Dunkirk area but including one at Cherbourg.

In the United Kingdom the Army had thirty-nine operational squadrons, of which two were in Ireland. The Navy had thirty-one. Twenty-nine of these have no role allotted, and are stationed at ports and anchorages around the coast. One of the two squadrons formed later and marked as non-operational is stationed at Machrihanish, and the Cherbourg squadron has no role allotted. We may therefore assume that these squadrons reflect the Navy's main preoccupation with submarine hunting in coastal waters and as being flying boat or sea-plane units.

Otherwise, the Army controlled four squadrons in Italy, one in Greece (at Salonika), three in Palestine, four in Mesopotamia, two in India, and four in Egypt, one of which was in East Africa, and one in the Sudan. The Navy had five squadrons in the Eastern Mediterranean, that is at bases like Imbros. And there are three bomber squadrons in France, one with a Naval number, which are noted as belonging to the Independent Air Force.

Squadrons are often, not always, marked with their roles. In France, the most frequent role is fighter, or night fighter, thirty-one altogether. There were seven day-bomber units, five night-bomber, fourteen observation (i.e. artillery spotting) squadrons, and three reconnaissance. Of the thirteen Naval squadrons altogether in France, one is the probable sea-plane or flying-boat unit at Cherbourg, six are positively marked as fighter, one as reconnaissance-bomber, three as bomber and two as night-bomber.

Otherwise, the Salonika squadron is marked as day-bomber. Italy had two fighter squadrons, one fighter-bomber, and one reconnaissance. The squadron in East Africa is marked as reconnaissance, and the other three squadrons in Egypt are bomber;

one was moved to Palestine before the end of the war. The Palestine squadrons are one fighter, one observation, and one bomber. In Mesopotamia, there were one bomber-transport and two fighter squadrons.

But the United Kingdom picture is more complex. There were one Observation and one Reconnaissance Squadron in Ireland. In this island, there were, from the RFC original strength, seventeen fighter squadrons, six night-fighter, nine bomber and five night-bomber. From the Naval side there were one fighter and two unattributed squadrons. The operational strength in the UK was obviously directed towards home defence. The soaking up of twenty-four squadrons and the shaking of an administration, must have made the otherwise moderate German bomber deployment, estimated to be at most forty aeroplanes, worthwhile.[2]

There are also in the United Kingdom sixteen operational squadrons which have no role allotted. It is likely that these were the Training Squadrons. The Official History states that there were nineteen Training Squadrons by the end of the war, and presumably these are included. There were also fifty-six Home Depot Stations, each housing three squadrons, and it is likely that here we have a confusion of thought between flying squadrons and handy-sized units of men. We know that there were a vast number of specialist postgraduate courses for trained pilots, but there is less information than one would imagine on the actual course of ab initio pilot training, which was, presumably, the task of the Training Squadrons. Most writers of memoirs talk about how they went solo and then skip straight to their squadrons, which must loom as more important in their memoirs.

Certainly training was unsatisfactory. The usual tale is of men going into action with only a few hours solo in their logbooks. Losses were blamed on bad aeroplanes, and many *were* due to that. However, by the middle of 1916 it had dawned on the High Command that not all losses need be blamed on the aeroplanes. Any aeroplane could be fatal if the pilot had not been properly trained, and almost any aeroplane could be useful if the pilot were used to flying, not as an occasional treat but as a job to be done every day, and if he were used to that machine, and if his sorties were properly allocated and planned to keep him out of harm's way. It was not till June 1916 that minimum standards were laid down; a man should not be sent out to France till he had completed at least fifteen hours solo, and had made a cross country flight during which he had found two other airfields and landed at them. He must also be able to climb to 6,000 feet, and then glide to a safe landing with engine off. He must have landed twice in the dark using a flare path, and been checked out on a service, that is an operational aeroplane. In other words it had been noticed that aeroplanes and men were being wasted simply because the pilots had got lost, and weren't able to find their way home from a sortie, often their first, and that when their fuel gave out or their engine broke down they weren't able to make a forced landing even on a prepared friendly airfield. And many of them were so unfamiliar with the air that an emergency like that just threw them.[3]

Training pilots was never a very safe or efficient operation till near the end of the war when Smith-Barry began to think about how to do it, and developed the Gosport system of effective dual instruction and the instructor's 'patters'.[4] Courage and simple skill in flying were not enough to make a flying instructor, and neither were battle fatigue and the need for a rest. Instructors had to be taught, and that became the task of the Central Flying School.

There were other training units in the United Kingdom. The training of pilots started with a ground phase at the Cadet Brigade at Shorncliffe; there was another such brigade in Egypt and one, possibly originally an Admiralty creation, in Canada. Officers with experience in ground operations were the preferred training material, although direct entrants were acceptable. Besides aeroplane pilot training, there was also a School for Balloonists at Larkhill and one at Lydd. There was even a School of Free Ballooning at Hurlingham, though where these balloons were expected to drift to uncontrolled in wartime remains a mystery. There was an Airship School at Cranwell, an old Naval Training Station.

The training of ground recruits was vastly complex. Boy Training was at Eastchurch, another old Naval Station. Early in the war Lord Rothschild had given his estate at Halton to the war effort. The Army had first used it as the RFC Recruits Training Centre, and it later housed the School of Technical

Training (Men). The Wireless Schools were at Flowerdown and at Penshurst. Armament was taught at Uxbridge, ground armament at Eastchurch. Farnborough housed a School of Photography. Other trades were catered for at schools which passed out of the Service's occupation after the war.

Between 1 April 1918 and the end of the war, nine more land-based squadrons were formed, and five coastal stations at home and six in the Mediterranean. At the end of the war, the strength was 27,333 officers, more than half of whom were pilots, and 263,827 other ranks. There were in total 22,847 aeroplanes on charge.

And in April, down in France, there were three squadrons allotted to the Independent Air Force. Independent of what?

Trenchard and the Independent Force

On 14 April, Trenchard resigned from his post as Chief of the Air Staff. Sykes took over. On 25 April, Rothermere also resigned: he had little option, it was not only Trenchard who had walked out but also Sir Godfrey Paine and Sir David Henderson, the Navy and Army Directors of Aviation, as well as some of his higher grade civilian temporary officials. Weir took over the Ministry.

Boyle paints a picture of Trenchard, unemployed, sitting on a park bench in the usual manner of the workless, having to be coaxed back to some kind of duty by Lord Weir. Weir's biographer is unable to accept this situation, but asks why Trenchard could not simply be ordered back to a post. This shows how little experience he has of generals and the like.[5]

The episodes of the Second World War have accustomed us to the spectacle of very senior officers behaving like prima donnas in public. But they *are* prima donnas, the leading actors in the theatre of war. If a man has for some years held high command, and seen all his orders obeyed, it is not to be wondered at if he sees the respect and deference due to his office as properly due to him personally. Just about this time, Beatty had been promoted to be Commander-in-Chief of the Grand Fleet from command of the Battle Cruiser Squadron, over the heads of two other battle-ship squadron commanders. The Lords of Admiralty were not brave enough to order these admirals to haul down their flags, that is to dismiss them, but they did suggest to them that they might prefer to haul down their flags of their own free will. One, Jerram, agreed and left the Fleet: the other, Sturdee, who at least had the excuse that he had actually won a battle, refused and sat it out in post for another fifteen months.[6]

Weir did not want to lose Trenchard to the new Air Force. He had to talk him out of his rage, not so much with Sykes as with Rothermere. He then had to find a command for him to be offered which he could accept with dignity and without loss of face. There had already been enough publicity over Trenchard's resignation, which had resulted in a full-scale Parliamentary debate. Rothermere had resigned on the day before the debate. To let Trenchard go altogether would present a most embarrassing public relations problem (not all the press was under Harmsworth control, and Beaverbrook was a formidable political figure), to say nothing of a morale problem within the new Air Force. There was no way of sacking any other junior commander for Trenchard to succeed him, it would be too transparent. But there was a possibility.

In the October of 1917, during the height of the political scare over the air raids, Lloyd George had ordered the Royal Flying Corps to undertake what can only be called reprisal raids. After the 1914 raids on the Zeppelin works at Friederichshafen, the RNAS had hankered after more action on the same lines. In the summer of 1916, they had moved a wing of aeroplanes down to an airfield in the French zone of operations, probably simply because it was out of the reach of the Army. From this base at Luxueil they proposed to carry out a series of raids into Germany. While they began to prepare, the Army pleaded with the Admiralty for more aeroplanes for use in the Battle of the Somme; they were given the first machines intended for Luxueil and as a result, the bombing campaign did not start till October. During the spring, the RFC again asked for help, and the squadrons from Luxueil were, one by one, lent to the RFC. By the early summer, the naval bombing had stopped. It was on 1 October 1917 that Trenchard was asked to release aeroplanes for operations from a separate base against the German homeland. The RFC therefore set up a base at Ochey, still within the French zone of operations, and moved down two of their own squadrons, one a night-bomber squadron equipped with FE2s, painted black. They also borrowed the RNAS 16 Squadron.

Trenchard appointed Lieutenant-Colonel Newall, on promotion, as commander of this VIII Wing.

Weir's plan was to involve Trenchard as commander of this detached bombing force. There was no real problem in getting Trenchard to take a field command with Sykes as Chief of Air Staff. Trenchard had had three years of observing Haig's relationships as Commander-in-Chief in France with a more junior officer, Robertson, as Chief of the Imperial General Staff. Haig certainly had not treated the CIGS as his superior, but rather as an agent to provide him with the manpower and the supplies which he needed, and to stand as a buffer between him and the politicians.

One of the problems was Salmond. John Salmond had been in command of the Training Division in Britain, as a Major-General, when he was brought over to take Trenchard's place in France, still as a Major-General. We must assume that it was Trenchard's decision to bring him over. Trenchard had to be persuaded to accept an appointment which could at least be plausibly presented as not subject to Salmond and the RAF Headquarters in France. It was unlikely that there would have been personal friction between Salmond and Trenchard; the relationship would probably have been parallel to that between the RSM at Sandhurst and the Gentlemen cadets when, according to legend, he told them that, 'You call me Mister, and I call you Mister, and you're the ones who mean it.' (Not, please note, 'Sir'.) But the Independent Air Force was independent because it was essential that Trenchard should be independent of Salmond, and be seen to be independent. Rothermere certainly did not see it that way. He writes gleefully of 'that dud Trenchard now being sent to France as second string to Salmond.'[7]

The other problem *was* Sykes, but not Sykes as a person. Rather it was Sykes' view of the Air Weapon. To Trenchard, the long service regular infantry officer, the Air Arm was essentially a part of the Army, just like the Artillery or the Engineers. It could have no independent role and therefore no independent existence. The units served as an integral part of the higher formations of the Army, usually as corps troops, although they might be deployed as low as divisional troops. And though there might be a central force, under the General Headquarters of the RFC, that was not an independent weapon to fight its own campaign, but rather a reserve to be put at the disposal of an Army commander for use in a great set-piece attack or some such other situation where maximum effort was being demanded. When Henderson left HQ RFC for his brief period as an infantry division commander, and Sykes became for a short period temporary commander of the Air Force in France, he began a process of withdrawing the Wings from the immediate authority of the Corps, or Armies, to which they had been attached, and tried to centralize them under the RFC Headquarters. They should not, under this system, receive their orders direct from Army with copy to RFC HQ for information, but Army should ask RFC HQ to issue the orders to the Air Units stationed in the Army area. In contrast Trenchard did not see air operations as a matter of strategy for central detailed control, but only of tactics.

Trenchard had to be persuaded that an Air Force could operate quite divorced from the activity of an Army, and carry on a war of its own against the enemy's manufacturing resources. Weir did his job, and persuaded Trenchard. It was not a case of the task making the man, but of the task being given to a man who, by coincidence, happened to be available and free, and if left unemployed a potential political embarrassment. The new job needed someone with the Trenchard touch, to progress-chase in the building up of a large Air Force with its own airfields and engineering back-up, workshops and stores, and to mould the crews into a family. If Trenchard himself were available and willing, there was no need to look for anyone else. The real problem, if there were one, would be with Brigadier-General Newall, who had worked all through the winter setting up the operation. If it were not possible to send Trenchard to appear to be under Salmond's command, then he must on no account be sent to replace an officer junior to him, certainly not one who had already done a good job. So Newall would have to be retained, even if two generals seemed a large allowance for a force of three squadrons. But there were plans for expansion.[8]

Trenchard was not without support, although whether he welcomed it is another matter. In June 1918 Sykes sent in a paper to the War Cabinet summing up a meeting of the Inter-Allied Aviation Committee. They had discussed a proposal for an allied bombing force for long-distance bombing. The Americans and the French wanted a force to be subordinate to the Supreme

War Commander, that is to Foch. Sykes as the British representative had argued that the force should be completely independent of the Armies, and should therefore be responsible solely to the Supreme War Council. Sykes pointed out that the importance that the British attached to their bombing campaign could be gauged by the fact that the most experienced British Air Commander, General Trenchard, had been selected to head it. He advocated that the new inter-allied force should be commanded by General Trenchard. It had been hard enough to get a British bombing force off the ground: getting an inter-allied force together staggers even the most imaginative. A decision in principle to do this was taken, at the end of October 1918.

Baring went to Ochey with Trenchard, to take up his old place. He was a major now. Where Baring had been while Trenchard was in England we do not know, but he was essential to the scheme of things, to act as Trenchard's memory. It was while Trenchard was at Ochey that a French pilot said something which so exactly expressed Trenchard's view that he repeated it at once to his driver, and to everyone he saw, afterwards. The Frenchman said: 'Aviation is not a sport, it is war.'

During August and September, five more squadrons, one of them ex-Navy, joined the independent force. The last to come, 45 Squadron, was equipped with Sopwith Camels, which, like FE2s, are not usually thought of as bombers. The FE2s were, it seems, normally employed as intruders attacking airfields. In fact, while the Force dropped 550 tons of bombs during its existence, it dropped nearly half of them, 220 tons, on enemy airfields. The Air Offensive as usual generated its own opposition and created air war for its own sake.

In England a fresh Group (the new RAF term for Brigade) was created to use the new big Handley Page aeroplanes to bomb Berlin. The operating base was to be in the United Kingdom, not in France, but there is some ambiguity as to whether this was to be part of the Independent Air Force. The aeroplanes were not ready till 8 November. By then the Turks had surrendered, and Austria had disintegrated. It was being proposed to move the bombers from Palestine and Mesopotamia to near Prague, where the distance to Berlin was only 150 miles and the attack would be feasible. But in early November, the war with Germany was over, and Haig's army of the County Regiments had reached the far frontiers of Belgium. The most attractive target of the lot, Berlin, faded from view.

The Independent Air Force stopped operations. It could pack up and go home. By the beginning of the new year, Trenchard was CAS again, Weir was on the way to Glasgow, Sykes was Controller General of Civil Aviation, and Churchill was wearing two hats. Churchill's dual post was the real menace. There would be no trouble, interested parties thought, about dissolving the union of the Military and Naval Air Forces. Trenchard had up to now been the advocate of the Air Force as part of the Army, and an opponent of the independent idea.

Dissolution

Neither Churchill nor the rest of the Government were in too much of a hurry to split the Air Force between the two older Ministries. The breaking down mechanism was enough of a problem without a reunion problem to go with it.

The disbanding dates of squadrons reminds us that what was signed on 9 November 1918, to become effective on 11 November, was merely an armistice, and not a treaty of peace. That came only with signature of the Treaty of Versailles on 28 June 1919. Of the RAF squadrons in France, only three were disbanded between the Armistice and the Treaty. But twelve were disbanded in June 1919, and eight in July. The big sweep came in December 1919 and January 1920, when thirty-seven squadrons were disbanded. After that there were hardly any squadrons left in France to be disbanded, although they included three of the Independent Air Force squadrons, which hung on till April 1920. Disbandment in the United Kingdom started earlier. Most of the Training Squadrons were gone by the end of August 1918, and the Air Defence Squadrons slipped away in ones and twos, through 1919, till in June 1919 ten of the old RFC Squadrons were disbanded at once. The Naval sub-hunting Squadrons followed the same pattern, with ten going in May 1919, and the last five in June.

The aeroplanes were being broken up and the men returned to civil life. We may here note what happened to several of the other men who played major parts, directly or indirectly, in the foundation of the Air Arm.

Haig and Robertson left the active Army. Field-Marshals never formally retire, they simply pass into unemployment and die in their beds: mostly. Field Marshal Sir Henry Wilson served out his term as Chief of the Imperial General Staff, and when he left the War Office went into Parliament: he was returned for Paignton, unopposed, in February 1922. One morning in June 1922 he left his house in uniform, because he was going to a memorial service. On his doorstep he was shot dead by an Irish extremist. The murderer knew he could not get away because he had a wooden leg.

Sir David Henderson resigned his post as Director-General of Military Aeronautics when Trenchard resigned as CAS. He refused to take a commission in the Royal Air Force. He left the Army at the end of the war, and went to Geneva as Director-General of the League of Red Cross Societies. He died in August 1921.

Maurice Baring left the Royal Air Force and returned to writing. He lists his clubs in *Who's Who*, and they may give some idea of the life he had left and returned to: they were White's, the Athenaeum, the Turf, Buck's and the Beefsteak. He was given the OBE at the end of the war, and was granted an honorary commission as a Wing Commander in the Reserve of Air Force Officers in 1925. A small enough return for his services. He died in 1935.

Major-General E.D. Swinton, Ole Luke Oie, who had been in trouble for his support of tanks, after a tour as an Assistant Secretary at the Committee of Imperial Defence, went to the Department of Civil Aviation as Director of Information, presumably under Sykes. In 1923, he was elected Chichele Professor of Military History at Oxford. The massive work-load imposed on professors at Oxbridge may be gauged from the fact that he virtually ceased to publish. He died in 1951.

And Major-General Sir Frederick Sykes? The Director-General of Civil Aviation continued in that post for another year, as well as in his membership of the British delegation to the Versailles Peace Conference. He had taken the Civil Aviation post willingly, because after all it was Sykes, and not Trenchard, who had the great Imperial dream. Again we do not know what he said when he first saw the estimates for the new Air Ministry. Out of a total expenditure of £21,471,495 anticipated for 1920–21, Civil Aviation was to get £894,540. Even so he hung on for another year, and then resigned. By now he had married the

Prime Minister's daughter. Trenchard got married at about the same time. In those days it simply was not the custom for professional soldiers to marry young.

In 1922, Sir Frederick Sykes retired from the Air Ministry, and entered Parliament, in the Conservative interest, sitting for the Hallam division of Sheffield: for this he turned down the Governorship of South Australia. His many questions on Air matters gave innocent employment for a number of civil servants. When in the following year the Government set up a Broadcasting Committee, Sykes was made Chairman. He saw through a report that there should be a public body to carry out broadcasting, and suggested an organization. The report was accepted, and he became the Chairman of the Broadcasting Board, and was instrumental in appointing John Reith as Director-General.

After his statutory three years at the BBC were up, Sir Frederick Sykes was appointed Governor of Bengal. He served the usual five years, then returned to England and busied himself with the numerous directorships which he had collected over the post-war years. In 1939, he offered his services, in uniform, but was not needed. He again entered Parliament, and in 1940 was returned unopposed, under the electoral truce, as Member for Central Nottingham. He died in 1951. He was a man of character and ability, who did the State much service. It would be wrong if he were entirely forgotten.

But Major-General Sir John Capper outlived them all. After a youth spent in small wars in India and Burma, and a middle age, in military terms, spent in balloons and clinging to Kites drawn by galloping horses, he commanded an infantry division in France. In 1917 he went to the War Office as Director-General of the Tank Corps. When the war was over he retired, but from 1923 to 1934 he still served as Colonel Commandant of the Royal Tank Corps, being succeeded in that honorary post by Swinton. He died in 1955, at the age of ninety-one, the last of the men whose work, before 1914, ensured that the concentration camp fences and the crematorium chimneys would not rise on English fields.

The Memoranda
Each of the first two Chiefs of the Air Staff settled down to plan

the future Air Force. Their thoughts were put on paper in two Memoranda. The philosopher-historian Collingwood remarked that everyone analyses Nelson's plan for the Battle of Trafalgar. Nobody analyses Villeneuve's plan for the battle, because Villeneuve lost. For the same reason, everyone writes about what is always called the Trenchard Memorandum, but nobody writes about the Sykes Memorandum which preceded it.

The Sykes paper is reprinted in his autobiography, as an Appendix. It is dated 9 December 1918. The title is significant, and in a way says it all: *A memorandum by the Chief of the Air Staff on the Air-power requirements of the Empire.* The document is quite long, and meticulously subdivided, a classic staff paper. It is concerned with all the air requirements, both military and civil, which should be supplied by the Air Ministry. This, the home in London of the Imperial Air Staff, must include representation by the commercial air interests.

Sykes envisaged an Empire held together by air power, that is by civil aviation. Where commercial firms do not provide the mail carrying and, perhaps, the passenger carrying capacity, the Royal Air Force must look after it. The size of the Imperial commercial air fleet and the powers of control to be exercised over it by the Government have yet to be decided, and make forecasting difficult. But Sykes comes down heavily for a large degree of State assistance and of control. The industry built up during the war, he says, will probably wither away if it is not supported by the State. 'The future of commercial aviation must depend largely on the Royal Air Force for the provision of the necessary pilots and technical workers'; therefore the training of Service pilots must include training to operate as commercial pilots also. The Government must use world-wide state propaganda – this was not yet a dirty word – in support of the British aircraft industry.

It is Sykes' military proposals which are most relevant here. A large part was played by the concept of the Cadre squadron, that is a squadron with a permanent cadre of a regular commander and flight commanders, and a group of ground crew, but mainly manned both in the air and on the ground by part-time volunteers.

Sykes divides aircraft for the United Kingdom and home waters into those required to co-operate with Sea Forces and with

Land Forces. For the Grand Fleet and Home Waters he envisages four squadrons of ships' fighters and fighter reconnaissance, and four of torpedo planes and ships' bombers, all, it appears, regular full-time squadrons, because he specifically adds four squadrons in cadre day bombers, five squadrons in cadre large flying boats, and six squadrons in cadre large day bombers. For these cadre squadrons, bases would be necessary in the Azores, in Newfoundland, and in the Gulf of Saint Lawrence. In the Mediterranean and the Pacific, he adds two squadrons ships' fighters and fighter reconnaissance, four squadrons torpedo planes and ships' bombers, three squadrons of large flying boats, two squadrons of day bombers and three of large day bombers. Thus for the Fleet and for coastal work, he allocates twenty-two regular squadrons and fifteen cadre squadrons.

For Home Defence he allocates cadre squadrons only, forty-one of fighters of different types, eight of day bombers and eight of night bombers, fifty-seven squadrons in all. In addition he advises that there should be a proportion of squadrons available for co-operation with tanks.

Overseas he allocates regular squadrons only. Between Egypt, Mesopotamia, India and East and West Africa, he plans for eight squadrons of fighters, four of reconnaissance and Army co-operation, four of day bombers, and six of large day bombers. Twenty-two squadrons in all. For the self-governing dominions, he advocates only cadre squadrons, twelve for Canada, ten for Australia, and six for New Zealand. For South Africa he allocates nine squadrons in cadre, together with a large and well-equipped aircraft depot at Cape Town.

What is impressive about this memorandum is that there is virtually no mention of finance. And out of sixteen pages of the printed memorandum, less than two pages are given over to manning. Sykes assumes that there will be compulsory military service, with boys doing four years cadet service followed by one year full-time service. Those who then wished to serve on might volunteer for an engagement of seven years regular, and five years reserve, which was the pre-war normal army term. If there were no conscription, the technical trades would be filled by boys, enlisting at age about fourteen for twelve years service and receiving three years technical training, and by men 'with suitable qualifications' enlisting for two years with the colours and ten on the reserve. Non-technical personnel would be enlisted at age 18-25, for eight years regular and four years reserve, and such as showed aptitude would receive technical training also. And apart from these lengths of engagement there is no mention of any time scale either. Both Technical Officers and Administrative and Staff Officers would be drawn from the ranks of 'aviation personnel who, from one cause or another, are unsuited for normal duty in the air'.

In fact there is nothing to suggest that there had been any detailed thinking through of the whole scheme. Could it have been sold to the politicians? Not in the immediate post-war period, the era of recovery from the war to end wars, a time of pacifism, not so much active or militant as depressed and disgusted. This vast scheme, for 154 squadrons ready for service in an emergency, would have seemed like a last blast of an outworn imperialism. But echoes of that blast continued to reverberate right down to the middle of the next war, as separate items of Sykes' scheme were found relevant and resurrected.

A year later, on 19 November 1919, the new Chief of the Air Staff sent on to the Secretary of State 'A Memorandum on The Permanent Royal Air Force'. This was issued as a White Paper on 11 December and published in *Flight* on 18 December. It was a shorter and very different paper. While the Sykes paper was written in the full flush of victory with expense no object, it looks as if Trenchard, or his staff, were told to work out the best they could do with £15,000,000 a year, and lucky if they got that. The strength given for 1920 was twenty-three squadrons and seven flights, at home and abroad. By the end of the financial year 1922-3, six of the detached flights would be turned into seven squadrons, almost all for work with the Fleet. The matter of the other units, schools and depots and experimental stations, seemed much more important. It was assumed that there would be no need for several years for anything resembling general mobilization. The stress throughout the paper was, therefore, on planning for the future.

There are two ways of looking at an Air Force. The first is to see it as existing only in the four minutes between the declaration of war and the arrival of the missiles. The other is to see it is as something permanent and stable, like the Church of England. Sykes was by nature a four-minute man: Trenchard was clearly a

Church of England partisan. But even the Church Militant, the poet tells us, although it looks so firm to us is only flesh and blood.

In contrast with Sykes, once the business of the number of squadrons and co-operation with the Navy has been got out of the way, most of the rest of Trenchard's paper is concerned with manning and training. Trenchard states the aims of his plan. The first consideration is to build up 'an Air Force Spirit', and it is for this reason that the permanent officers must be trained at an Air Force Cadet College, not by the Army or the Navy, which Sykes would have been satisfied with. The second consideration is to reduce the number of flying accidents by training the mechanics, 'recruiting the bulk of the ground staff as boys and training them ourselves'. The third consideration was that it was not sufficient to make the Air Force officer a chauffeur and nothing more. The development of the independent service under Trenchard, effectively till his retirement in 1930, is a process of fulfilling these considerations, while drifting back to his original picture of the RAF as a tactical weapon of the Army or the Navy as required.

Publications

The Air Force had begun, and the regular operation depended in a large way on the production of two periodicals. One was the *Air Force Estimates* and the *White Paper* accompanying and explaining it. The other was the monthly *Air Force List*. Both were issued free to those members of the organization who needed them, and were also on sale to the general public. An answer to a Parliamentary Question about the *List* in 1923 says that 2,000 copies were printed each month, of which 675 were sold.[11]

The *Estimates* were just that, the best estimate the Ministry could produce of how much it would cost to run the Service for the coming year, which it was asking Parliament to provide. The *Estimates* as published are therefore a good source, in that they are meant as a basis for action: the *White Paper* is a rather less good source, since it represents the Ministry's attempt to justify the estimates and present the official line.

The *Estimates* are presented as a number of separate numbered Votes. Thus Vote 1 deals with the pay and allowances of the Air Force, Vote 3 with warlike stores like airframes, engines, armaments, and fuel, Vote 7 with half pay and pensions, Vote 5 with the cost of the Air Ministry, and Vote 8 with Civil Aviation (Tables 5, 6 and 7). They also include detailed explanations of some items of expenditure; for example the early lists are much concerned with the transfer of stores from the Admiralty and the War Office, and with other matters of inter-departmental accounting, as well as with the disposal of stores left in the Irish Free State.

There are also detailed accounts of losses which cannot be recovered. In the earliest Lists these include a number of cases of losses of actual, physical, money which cannot be recovered. There are two quite regular types of case. In one type, an officer draws public funds from a bank, for example the pay for a station or for the Mess Staff, in notes or coin, at least in some easily convertible form. For this purpose he is taken to and from the bank in a Service vehicle. On the return journey the officer stops the vehicle so that he and the driver can go for coffee, or for some other necessary purpose. They do this out of sight of the vehicle, in which the money is left. When they finish their diversion and go back to the vehicle, the money is gone. In some cases, where they have left the keys in the vehicle, or even left the engine running, the vehicle is gone also. In none of the cases described has the money been recovered; sometimes even the vehicle is gone for good.

In the other regular type of occurrence, an officer withdraws in cash a large sum in public or non-public funds; for example, he asks to be given the entire cash holdings of the mess, and has sufficient authority for this to be done. He then leaves this money, in notes or in coin, on his bedside table while he goes to dinner. As it is a warm evening, especially if he is in Egypt, he leaves his bedroom window open. When he returns, the money has disappeared, though no one else has been robbed. Again, in none of the cases described has the thief been caught or the money recovered.

The *Estimates* then, are a *good* source. The other regular publication, the monthly *Air Force List*, is a *very* good source. The *Air Force List* today, published half-yearly and available to anyone who wishes to pay for it across the counter of the nearest branch of Her Majesty's Stationery Office, is merely a list of the names of all the officers in the Service, arranged by branches in order of seniority. It contained separate Lists of the officers, including the senior civil servants, employed in each department at the Ministry and in the headquarters of the larger formations, such as Groups and Commands. Each List is in order of rank, and of seniority in each rank.

The *Air Force Lists* of the interwar period were much more interesting, both to the historian of today and to the spy of the time. They did include the seniority *Lists* of the branches, but they also included very much more. They listed every station in the Service, and every unit, every squadron and detached flight, every school. For each the names of all the officers employed were given, and the number, though not the names of the NCO pilots; the type of aircraft used by the unit was usually given (Tables, 7, 8, 9, 10). The *Lists* were not issued for the sake of the general public nor for the sake of the politicians or the financiers, although all might read them with profit. They were issued so that every officer could see where exactly he stood on the seniority lists, and therefore how near he might think himself to promotion.[9] He might turn to the end and see what had happened to the officers who had entered with him, whether they had, since he last looked, been killed, or invalided out, or transferred to the reserve, or cashiered or dismissed the service. And the *Lists* were issued so that officers could be found, so that, for example, their tailors in London or Lincoln could tell where to send their bills.

The *Lists* were fairly accurate. The *List* for any month is given as correct for the 15th of the preceding month, although it is possible, if you search, to find small inaccuracies, usually due to delays in the post. Nevertheless, they make it possible, for the 15th of any particular month, to count with a high degree of accuracy just how many pilots there were in the Service, how they were distributed geographically, what aeroplanes they flew, and how near each man is either to the beginning or the end of his tour. It does not appear, to the best of my reading, that any historian has made systematic use of this material. This material from the *Estimates* and from the *Lists* is the basis of what follows.

Ranks
The means whereby seniority and efficiency were reconciled in

promotions was a combination of qualifying Promotion Examinations and a device known as the Promotion Zone. In the original Scientific Corps, promotion had been strictly by seniority, and the same rule had applied to Naval officers above the rank of captain. It worked well enough in a static force, and resulted in a steady increase in the number of grey-haired lieutenants. In the RAF, promotion from pilot officer to flying officer was automatic after two years commissioned service, provided the man had passed his promotion examinations. In the same way, promotion to flight lieutenant was automatic after a further four years service, assuming the examinations had been passed. The squadron leader might not begin to hope for promotion on his merits, that is on the evidence of his annual confidential reports, till after he had been in that rank for a set period, which in the 1930s was three years (in the nature of things it cannot have varied much through the period); he had entered the Zone. If he still had not been promoted to wing commander after eight years as a squadron leader, he could forget about it; he had passed out of the Zone and would remain a squadron leader for the rest of his career.

The promotion Zone for the step from wing commander to group captain was between three and seven years in rank. From group captain to air commodore the Zone was from two to four years in rank, and for the next step to air vice-marshal was from two or four years as an air commodore.

These references are to permanent or substantive promotions, which are shown in the *Lists*. Acting, that is temporary, promotions, were not as a rule shown in the *Lists*. Today, these permanent promotions are dated from the days of the two half-yearly promotion lists, issued at New Year and in midsummer; experience suggests that it is unusual today for a man to be promoted into a rank which he does not already hold as an acting appointment, and delay in thus confirming an acting promotion is popularly seen as a black mark on the record; even for air marshals . . .

Service folklore is that the future Chief of the Air Staff has already been selected by the time he is a flight lieutenant, as in order to reach four stars before his retirement date he cannot be allowed to spend more than the minimum time at each junior rank. If a young and bright group captain flies into a mountain, it may cause a great deal of frenzied work for the Manning and Careers staff.

The rank titles taken for the new RAF and adopted on 4 August 1919, were drawn from those of the Royal Naval Air Service (Table 3). Under that heading in the *Navy List* of 1918, we find two groups of officers. One group, under the heading of RN officers, is shown as having ranks in the Royal Navy: The other group is shown under the heading of RNAS officers, and as holding temporary ranks: it must be assumed that these are officers who had joined the RNAS as Direct Entrants, not Naval officers who had transferred. All these officers are shown as having alternative titles, derived from their posts, which are not always easy to reconcile with their Naval ranks, whether substantive or temporary. These titles are those of flight or observer sub-lieutenants for the Naval officers, and of probationary flight or observer officers for the RNAS list. Above this were flight or observer lieutenants, flight or observer commanders, squadron commanders or squadron observers, wing commanders, and wing captains. The rank the Navy gave Colonel Sykes was that of temporary wing captain, RNAS. The post titles were shown by constellations of discreet gold stars on the sleeve, above the executive curl.

The new RAF titles were originally highly functional. For most of the 1920s a squadron was commanded by a squadron leader. The squadron usually had four flight lieutenants, who were to fill three flight commander posts, and one post for a qualified engineer (who was, of course, primarily employed as a pilot). The mass of the squadron pilots were flying officers or sergeants. A squadron usually had two or three pilot officers, who were very definitely under training: most pilot officers of the General Duties Branch were to be found at the flying training schools. By the 1930s the commanding officer posts for the heavy or night bomber squadrons were upgraded to be filled by wing commanders: these squadrons were organized as two flights, with squadron leaders as flight commanders.

Otherwise, wing commanders commanded wings of two squadrons, and also the stations at which their wings were based. A group captain was seen mainly as holding a staff appointment or as the commander of a major base: for a period the Commander-in-Chief of Aden Command was a group captain post. Otherwise the higher formations were commanded by Air Officers in a manner decided at the time on pragmatic grounds; it should be noted that air commodore has always been very much a junior Air Officer, while today the brigadier is to the Army what he was in the beginning, a senior form of colonel. The present normal pattern of Stations commanded by group captains reporting to air vice-marshals commanding groups was a product of the Second World War. It is in fact a puzzle as to why, with the lower ranks and titles being so functional, the early Air Council thought it necessary to create so many Air Ranks, as many as the Army had. It did in the end allow expansion on a rational pattern, but that was not obvious at the time. It was necessary to keep up with the Army if the RAF was going to be any use to the Army, or the nation: Brigadier-General Henderson had to go to France in command of the RAF in 1914 because Lt.-Colonel Sykes was not high enough in rank to argue with the generals, and similarly in later years the generals and the admirals were not going to listen to anyone who was not of Flag or General Officer rank. We may see Trenchard's aggressive behaviour towards Beatty as pressing home this point.

Officers of the rank of flight lieutenant and below are Junior Officers. Officers from squadron leader to group captain are Senior Officers. Officers of the rank of air commodore and above are not Senior Officers and should not be described as such: they are Air Officers. One may say that officers above captain RN/colonel/group captain level are important in the Service. Officers of the level of vice-admiral/lieutenant-general/air marshal (three-star officers in American usage) are at least potentially important in the State, and to show this they are normally in the British Service knighted on substantive promotion.

The Start Line

The first *Air Estimates* were for the year 1919-20, and were issued on 13 December 1918, that is, on the day that the Trenchard Memorandum was also published. These *Estimates* were for £54,030,850, which was less than the forecast *Estimates* for £66,500,000 issued in March 1919, the usual time for presenting *Estimates*. There had been a reduction in the amounts forecast for supply of aircraft and engines, and for land and building. In contrast, the demobilization of officers had gone on more slowly than was expected. There were also small wars still being waged in India, Egypt, North Russia and the Baltic. Most of the *Estimates* in fact covered a carryover from the war. The maximum strength during the year had been 150,000: by the beginning of the next financial year it was hoped that the strength would be down to 35,000 men. The *Estimates* for 1920 to 1921 are the first realistic *Estimates* of the Trenchard era. They cover twelve Air Officers in the field and seven at the Ministry, 2,880 Officers in the field, 144 at the Ministry, six employed on civil aviation, and nine employed on research and development. There were 152 Cadets, that is Officer Cadets. Other ranks totalled 324 Warrant Officers, 2,900 NCO's, 119,760 Airmen, and 3,470 Boys. There were also forty-three Other Ranks employed at the Ministry and twelve on Civil Aviation.

Officers' pay came to £1,409,000 and pay for other ranks to £1,558,000. Quartering estimates of £2,000,000 included a token £1,000 for forage. Aeroplanes and their engines and spares were to cost £1,389,750, and airships and all their engines and spares were to cost only £37,000. The airship force inherited from the Navy was being run down, and the later airship development ending in the R100 and R101 was financially a civil concern. The year's estimate for ammunition was £72,250. Works and buildings were to cost £3,647,000 and the biggest single vote after Vote 1, Pay. Vote 9, Experimental and Research, is for £2,575,540, with an additional £175,000 for the Directorate of Supply and Research at the Ministry. And Civil Aviation, Vote 8, was to get £894,540, with added to it £77,269 for the Meteorological Office, taken over from the Board of Trade, and £109,000 for the Department of Civil Aviation at the Ministry. Sykes' department was a minor consideration.

Notes and References

1. Pepys *Diary*. Pepys takes up this theme on 20 July 1662, and much of his diary is a commentary on it.
2. S/Ldr. J. Slessor, 'The Development of the RAF', *RUSI Journal* 1931.
3. AP125, 1936 edition, p. 154.
4. F.D. Tredrey, *Pioneer Pilot: the Great Smith Barry who taught the World to Fly*, Peter Davies, 1976.

5. Weir's biographer: W.J. Reader, *Architect of Air Power*, Collins, 1968.

6. A.J. Marder, *From the Dreadnought to Scapa Flow*, OUP, Vol. III, 1966, p. 287.

7. Reginald Pound and Geoffrey Harmsworth, *Northcliffe*, Cassell, 1959, p. 638, quotes a letter from Rothermere to Northcliffe dated Whit Sunday 1918.

8. Neville Jones, *The Origins of Strategic Bombing*, William Kimber, 1973, is a classic of military history, dealing in detail with both the political and the technical problems of the Independent Air Force.

9. David Niven, *The Moon's a Balloon*, Hamish Hamilton, 1971, deals with great perception with the life of the professional junior officer in the 1930s and with the classic system of the enclosed world of the traditional County Regiment. The *List* figures on p. 77.

Chapter 5

Squadrons and Higher Powers

The Initial Question

'Give me,' cried Galland, 'a squadron of Spitfires!' Why a squadron? Why not a Spitfire, or a hundred or a thousand, seeing that it was merely a rhetorical demand? Why any specific quantity, or what looks like a quantity? What was a squadron?

Of course, the answer to the question of how many aeroplanes is a squadron is like that to the question of how long is a piece of string. It is any quantity which is convenient to you at the time, and we shall see that that is a precise answer to the first question. A squadron is a unit, and when Sykes went about inventing most of the details of the new Flying Corps, the size of the unit was one of his first problems. He looked at the French *Escadrille*, which was about a platoon size unit with four or perhaps six aeroplanes, and at the German *Geschwader*, which was a battalion size unit with about forty to fifty aeroplanes. He split the difference. A major could be trusted to administer with little interference a unit of about the size of an infantry company or a field battery, that is something of about 200 to 250 men. That would support about twelve aeroplanes, with a few spares, and would require twelve pilots plus a few spares. Aeroplanes get broken or become unserviceable, and in the same way officers go on leave, or go on courses, or break their legs while breaking the aeroplanes, or go sick for non-service reasons and therefore at their own expense.

The squadron was therefore of a convenient size for subdividing into three smaller formations, of about four aeroplanes each, and therefore about four to six pilots, one of whom would be a flight commander; it would be a convenient post for a captain. The flight would be capable of operating for a short time detached from the main body. It would have a sergeant, a clerk to write down the hours flown, a cook who could keep the men from actually starving, a driver with a tender, and enough men to hoist the windsock, push the aeroplanes about and carry out the turnround procedures, change burst tyres and do minor tuning to engines or small bits of patching and joining on fabric and wooden frames. But these detachments would only be temporary. For any more fundamental repairs the aeroplanes would have to return to the squadron, which would have a much larger number of more skilled men and vehicles and more able clerks and cooks, and a larger stock of spares, and all the equipment and machine tools required for major repairs – even, as we shall see, for completely rebuilding an aeroplane. The squadron should be able to do anything that did not actually require returning the machine to its maker, via an aircraft park, or returning a man for major surgery at a real hospital.

It would seem that Sykes at the beginning, and certainly Trenchard after him, had a vague idea that a squadron would be a real and permanent entity like a county regiment to which a man, or an officer, could belong for at least a large proportion of his time in the service, if not permanently. There is no knowing what the squadron might not have developed into if the First World War had not interfered with evolution. However the squadron turned into a different kind of military unit, a parallel to the artillery battery rather than to the regiment.

The Army works on the idea of the regiment as a permanent unit, that is the man is a part of his regiment for the whole of his military career. It may be difficult to apply an ideal based on the County Regiment to something as large as the Royal Regiment of Artillery, but that is the basis of the Army's social thinking. It has proved very effective.

The Royal Navy works on the basis of a temporary unit. The theory still is that their Lordships take a ship, and out of a large pool of officers and ratings select a sufficient number of men of each relevant speciality to fill every seat in the ship. They put the men in, screw the top on, throw the ship into the water, and after three years, the length of the theoretical commission, bring the ship back and take the men out, to add them to the large pool from which they were originally taken. There is no reason why any two of them, let alone a larger number, should serve together in the same ship again while their service in the Navy lasts. But the ship lasts a lot longer than the men.

The squadron is an episodic system. Men join a squadron at random intervals to replace others, either because the others are dead or incapacitated, or because they have been routinely posted out at the end of their allotted time. This explains the occasional horror story of men arriving at squadrons in the middle of the Battle of Britain with very little experience on type straight from pilot training: while one department at Command or Air Ministry was filling casualty vacancies on squadrons engaged by posting in experienced pilots from squadrons in the unengaged Groups, another department was keeping up the normal flow of new pilots into squadrons direct from training units. The squadron is an episode in the life of the serviceman posted into it and later posted out: the man is an episode in the life of the squadron which existed before he came and will exist after he leaves. The squadron then is more a matter of a flowing and constantly changing society than is either the regiment or the ship's company.[1]

A squadron is normally composed of aircraft of a particular role, such as fighters. Throughout this and later chapters squadrons will be referred to as, say, 17F for a fighter squadron, or 18B for a bomber squadron, and so on for other specialities. In political statements on strength of the service throughout the 1920s and 1930s it is usual to see the number of squadrons quoted, with no further qualification, although sometimes the main types are distinguished. Increases or decreases in strength of an Air Force may be achieved either by increasing, or decreasing, the number of squadrons, or alternatively by keeping the number of squadrons the same but increasing or decreasing the number of aircraft on the establishment. Thus in the original Sykes concept, the squadron should consist of twelve aeroplanes: by the end of the war the establishment of the normal fighter squadron was twenty-four aeroplanes. This establishment went back to twelve as soon as peace broke out.

There were during the period some squadrons to which these provisions did not apply. For most of the period 24Comm.

Squadron was based at Hendon, and acted as a communications unit for officers based at the Air Ministry and for VIP transport. It was also used for training and qualifying by London-based officers who had to achieve a certain amount of flying every year to qualify for their flying pay. The types of aeroplane on strength varied greatly, and were a mixture of light transport and passenger machines, as the task demanded. It is never quite clear in political statements if 24Comm. is or is not counted as a front-line squadron, although it is difficult to see the Hawker Tomtit as a military weapon.

The Aircraft and Armament Establishment at Martlesham Heath was organized into two squadrons, one to deal with aircraft, and one with armament, testing and development. These two units, which were entirely local organizations for convenience of station administration, were each given a number as 15B and 22B.[3] Again, there is not always clarity in political statements whether these were being counted as active bomber squadrons. And at Aden was based a unit known as 8B, which at different times had a variety of aircraft, often several types with different roles at once, and for some time even had a fighter flight. But 8B is shown in the list like all the other 'genuine' operational squadrons, as consisting of a Headquarters and three Flights.

A squadron was just that, a headquarters and a number of flights. Not simply a given number of aeroplanes, or of men, but a node in the communications system, to which orders could be passed down and from which information could be passed up the chain of command. A squadron could be given an operation to carry out, a task to do, and could then look after the details itself. A Staff building up a strategic plan must do it in terms of squadrons.

The Changing Situation

In 1922, when the dust of demobilization and reconstruction had settled, the 'Striking Force' in the United Kingdom consisted of two fighter squadrons (25F and 100F) and two bomber squadrons (29B and 207B). Also in the United Kingdom were one Army co-operation squadron (2AC), one reconnaissance squadron for work with the Navy (203F), two flying boat squadrons (205FB and 230FB) and a squadron of 'ship's fighters' at Gosport (3F), together with a torpedo squadron (210TB). 4 photography squadron was stationed at Farnborough, and it is not clear whether we should count it as an operational squadron; 24Comm. was then at Kenley and will not be counted as operational. This makes, if we include 4Photo, eleven squadrons in all (Table 9).

Overseas, there were in Egypt, according to the *List*, 47B, 56F and 216Transport. The statement in the Trenchard Memorandum that there would be seven squadrons in Egypt by 1922 presumably also counts as in the Egyptian force the squadrons in Palestine, which are 14AC and 208AC. The *List* gives for Iraq 6AC, 8B, 30B, 45B, 55B, 70B, 84B, against a forecast of three squadrons for Mesopotamia.

Presumably we see here the effect of the decision to govern Iraq by the use of air power. This was not a new bright idea which had to be forced on the Government, untried. The Royal Flying Corps had already by 1920 carried out both in India and in Iraq several operations of the general type foreseen: communications, reconnaissance, bombing of villages, breaking up of bodies of nomads.[4] What the politicians and the soldiers, with bitter memories of the last phase of the Boer War, knew well, was that if the United Kingdom accepted the League of Nations mandate to govern Iraq, the country could be governed in no other way. Inter-service friction was not over the choice of means or even over the size of the effort required, although these were the factors mentioned in open debate: it was over control and the consequent number of posts, especially senior and staff officer posts as well as air or general officer posts, created and available for officers needing jobs to be posted or promoted into.

In India, we find at this time, 5AC, 20AC, 27B, 31AC and 60B, against the seven squadrons forecast in the Memorandum. The total number of squadrons overseas, including a flying boat squadron in the Mediterranean, is therefore eighteen, plus a flight at Aden (Table 9).

The *Air Estimates* for 1921-2 announce that five new squadrons are to be formed, while the *Estimates* for 1922-3 say that there will be a cut of two squadrons. The Geddes recommendation had been for a cut of eight-and-a-half squadrons. The *Estimates* for 1923-4 forecast that fifteen new squadrons were to be formed for Home Defence by April 1925, although only seven of these were to be formed wholly or partly during the financial year 1923-4. One of these was 99B squadron, formed specifically to crew and operate the new Avro Aldershot, called in the press a 'Giant Bomber', sufficiently large and important enough for each machine to be given a name like a warship.[5] It had one engine and weighed 10,000 lb take-off weight, about the same as the Sidestrand, later praised as a light and manoeuvrable machine.

The next *Estimates*, for 1924-5, say that eight new squadrons were to be formed to bring the total up to eighteen by April 1925. When we consult the *List* for May 1925, we find that there are ten fighter squadrons in the United Kingdom – 1F, 3F, 17F, 19F, 25F, 29F, 32F, 41F, 56F, and 111F. There are eight operational bomber squadrons, 9B, 11B, 12B, 39B, 58B, 99B, 100B and 207B. 15B and 22B at Martlesham have now appeared in the *List*. This presumably makes up the eighteen Home Defence squadrons. But also in the United Kingdom were four Army co-operation squadrons, 1AC, 4AC, 13AC, 16AC. The naval needs were met by a number of flights, some landbased, some to be embarked when necessary in the Fleet, and will be considered separately.

The *Estimates* for 1925 claimed that the strength was now equivalent to fifty-four squadrons, being forty-five and two-thirds actual squadrons and twenty-one flights averaging six aircraft each. A count through the actual squadrons on the *List* for May 1925 shows only forty-one regular squadrons counted by numberplates, and that includes the communication squadron still at Kenley, and the Martlesham Heath squadrons. There were not yet any reserve or auxiliary squadrons. There are five landbased UK flights, including a 'Meteorological Flight' at Duxford, and a 'Night Flying Flight' at Biggin Hill. The Navy has attached twenty flights, including three Fairey IBIS flights, two on wheels and one on floats, at Malta.

But these estimates look forward. It is proposed that during the coming year two new regular squadrons will be formed, as well as one special reserve and four Auxiliary, which appear for the first time.

The special reserve squadron deserves some comment. This squadron was in fact a cadre squadron of the type which Sykes saw as providing the defence of the United Kingdom. Several of these squadrons were eventually formed, and can be distinguished by the numbers which were of the 500 series. 502 Ulster Special Reserve, the first cadre squadron, was armed with the Vickers Vimy, an aeroplane which had seen its best days at the end of the war, and was now being withdrawn from the last regular squadron to use it.[6]

The special reserve squadrons were in some measure the equivalent of the territorial battalions of the County Regiments, or even territorial units of the Corps. In contrast the Auxiliary squadrons resemble much more closely the old Yeomanry regiments, drawn from the landed and fairly well-to-do gentry of the countryside. The special reserve squadrons were mostly organized as heavy bomber squadrons, while the Auxiliary squadrons, with their air of the gentlemens' club or the Hunt, were day or light bomber units. The five Auxiliary squadrons in service by May 1927 were, like the cadre squadrons, equipped with aeroplanes no longer in use for any other purpose. The five Auxiliary squadrons in service by May 1927 were given DH9as, but were during the next two years re-armed with the Wapiti, a general purpose aircraft originally bought as a replacement for the DH9a, to use up the large stocks of DH9a spares and components which the Service held. Not all of the original Auxiliary squadrons were, in the end, mobilized as bomber squadrons. But late in the 1930s the special reserve squadrons were converted, perhaps promoted, into Auxiliary squadrons. Enthusiastic script writers, trying to mention everybody, may note that there were never any Volunteer Reserve Squadrons on the List.

The *Air Estimates* for 1926 admit that 'the original programme for a Home Defence Force of fifty-two squadrons will not be completed in 1928 as planned. The very complicated interlocking programme of land purchase, building, recruiting and equipment could hardly be completed before the year 1930.' This is the first hint in these papers that an air force could not be increased at the stroke of a pen. The fifty-two squadrons should include thirty-nine regular: it therefore appears that it was intended to form a total of thirteen Auxiliary or special reserve squadrons by 1930. In cold fact, these thirteen non-regular squadrons could at that time only be regarded as operationally available by the most sanguine of optimists. They could only be seen as suppliers of reserves against casualties, available some weeks after the opening of hostilities. Trenchard, in a speech to Cambridge University Air Squadron in 1927 reported in *Flight*, said that he expected a casualty rate of 70 per cent in pilots during the first week of a major war. The reserve crews would need conversion training on to operational types: the outworn machines with which they were familiar would not stand up to use in a real war. Nevertheless, the claim to have an Air Defence Force of fifty-two squadrons would be good enough for the public, or even for other politicians. But it is very difficult to reconcile the claims of the *Estimates* and Parliamentary statements with the number of squadrons shown in the *Lists* for each May, that is correct at Budget Day.

Trenchard retired at the end of 1930, and therefore the *List* for May 1931 may be taken to represent the state of the Home Defence Force he left behind. In the United Kingdom there were thirteen operational fighter squadrons, twelve conventional bomber squadrons, and one landbased torpedo squadron. Of the bomber squadrons, seven, with Virginia (4), Hinaidi (2) and Sidestrand (1) may be counted as heavy bombers for the time. They were each commanded by a wing commander, and were organized into two flights. The other bomber squadrons, with Fairey IIIF (2), Gordon (2) and Hart (1), were like the fighter squadrons organized into three flights apiece, and commanded by squadron leaders. There were five Army co-operation squadrons, each with the Atlas, and three flying boat squadrons, two at Mount Batten and one at Calshot. Thus twenty-five regular squadrons were available for a conventional air war in defence of the Kingdom, with the personnel of four cadre squadrons, and seven Auxiliary squadrons – one just forming – in reserve. But it might be claimed, throwing in 24Comm. and the two titular bomber squadrons at Martlesham Heath, that there existed thirty-seven regular and eleven non-regular squadrons for the defence of the United Kingdom.

There were no fighter squadrons abroad. There were bomber squadrons at Aden, in Mesopotamia (four, including one bomber-transport), three in Egypt and four in India. There were army co-operation squadrons in India (4) and in Egypt (1). There were also flying boat squadrons in Basra and in Singapore, and a nominal flying boat squadron with Fairey IIIF float planes at Malta. This adds up to twenty squadrons abroad.

Of the units attached to the Navy, almost all were in carriers. We find eight fleet fighter flights, eleven fleet spotter and reconnaissance flights, (one at Gosport and two at Lee-on-Solent) and seven fleet torpedo bomber flights (two at Gosport). It we take three flights as equivalent to one squadron, then nearly nine squadrons were at their Lordships' disposal. It might then be claimed, and was when the financiers were around, that the service had sixty-six regular and eleven non-regular squadrons in commission.

However, the sceptic might counter-claim, and with justification, that the greater part of this array was, in fact, engaged not on the work that the Military Wing had been created to do, but on work first given to, or later developed by, the Royal Navy. The Military Wing would only have claimed the Army co-operation squadrons, and perhaps enough fighters to protect them at work; possibly it would have taken also the three day bomber squadrons. The rest of the Service, the heavy bombers and the flying boats and torpedo bombers, the fighter squadrons positioned for the defence of London, and the shipborne aircraft, had all originally been part of the task given to or developed by the Naval Wing. The irony of this situation, in view of the Army origins of Trenchard and most of the Air Officers of the time, and of the later well known disputes with the Navy who through the 1920s kept on asking for their aeroplanes back, needs no stressing.

The Stations

Today, it is the station, not the squadron, which is effectively the unit to which a man belongs. There are two points here. First, the squadron did not consist simply of the pilots who flew, mostly officers with a very few sergeants. The squadron was an organization of all ranks and trades, including the men who did the servicing of the aeroplanes, the storekeepers, the cooks and the drivers. There was a squadron headquarters, an adjutant, and an orderly room, just as in an Army unit. And the link between the squadron commander and the men was the squadron sergeant-major. It was a long time before the title was changed and he became a warrant officer.

Secondly, the distinction between the squadron and the station was either very real, or nonexistent, depending on circumstances. In general terms the rule was that where there was only one flying squadron on an airfield, the squadron commander, usually a squadron leader, was also the station commander; no separate station headquarters is recorded in the *List*. If there were two or more squadrons, there was also a separate station headquarters, headed by a wing commander.

In the May list of 1931, we find the heavy bomber squadrons at Worthy Down (7B and 58B, with Virginias although no separate station headquarters is listed), Boscombe Down (10B, Hinaidi), Upper Heyford (99B, Hinaidi, shared with 40B, Gordons) and Andover (101B, Sidestrand, shared with 12B, Harts.) Although Andover, Boscombe and Upper Heyford each have a station headquarters listed, with two or three officers on station strength, the highest rank is one squadron leader at Andover, so we must assume that the senior squadron commander would act as station commander. The fighter squadrons were at Tangmere (1F and 43F, both Siskins), Upavon (3F, 17F, both Bulldogs), Duxford (19F, Siskin), Kenley (23F, Gamecock, 32F, Bulldog), Hawkinge (25F, Siskin), North Weald (29F, 56F, Siskin), Northolt (41F, Siskin) and Hornchurch (54F and 111F, both with Bulldogs.)

Higher Formations
To avoid a blockage in the information lines, it was necessary that there should be other formations to act as nodes in the communications network between the Air Staff and the squadrons. During the World War, these formations had developed first as wings of more than one squadron operating from one airfield, and later of brigades at a yet higher level. In France the brigade was part of an Army system of authority and communication, and commanded all the squadrons, whether observation, reconnaissance, fighter, bomber or balloon, operating in that Army's area. In the United Kingdom, there were Home Defence and Marine Operational Brigades, and also Home (Training) Brigades.

By March 1920, the formations answering directly to the Chief of Air Staff in the United Kingdom were Northern Area, Southern Area, No. 11 (Irish) Group, Coastal Area, RAF Cranwell, RAF HQ Halton. In the following month, Northern and Southern areas were united in Inland Area.

It is important to note that these were merely geographical divisions. In May 1922, Headquarters Inland Area at Uxbridge controlled three groups and a separate wing. 1 Group at Kenley commanded 24F at Kenley and 25F at Hawkinge, 2FTS at Duxford, the Artillery and Gunnery School at Eastchurch, the Signals Co-operation Flight at Biggin Hill, the School of Technical Training (Men) at Manston, 1 Stores Depot at Kidbrooke, 4 Stores Depot at Ruislip, the Record Office at Ruislip, and the Packing Depot at Ascot. 7 Group at Andover commanded 4 Photo, the School of Photography and the Royal Aircraft Establishment all at Farnborough, 1FTS at Netheravon and 5FTS at Shotwick, the School of Army Co-operation at Old Sarum, 3 Stores Depot at Milton, the Electrical and Wireless School at Flowerdown, CFS at Upavon and the Staff College at Andover. From Spittlegate (so spelt for many years) HQ 11 Group commanded 39B and 100B at Spittlegate, 25F at Digby, and 207B at Bircham Newton. Area itself commanded direct Uxbridge, Northolt, the hospital at Finchley, the Central Medical Board, the Aircraft Depot at Henlow, and the MT Repair Depot at Shrewsbury.

It is difficult to see any logic in these arrangements, except the logic of location and not too much of that. One hopes for better things from the Coastal Area Headquarters at Tavistock Place in London, which controlled directly the Inspectorate of Recruiting, the Air Ministry Wireless, the Marine and Armament Establishment on the Isle of Grain, the Aircraft Experimental Establishment at Martlesham Heath, the Aircraft Depot at Donnibristle, and RAF Leuchars, which in turn controlled 203FB and 205FB, and the aircraft in *Argus*, *Eagle* and *Hermes*. 11 (Irish) Group had long disappeared from the list, and no regular squadrons were now stationed in Northern Ireland.

And that, roughly speaking, was all the Air Force, at least the Air Force in Britain. The organization suggests that the units in the United Kingdom were not seen as having any particular war role, rather that the Air Force in Britain was a training force, with a few squadrons parked at home, to give a variety of posts for men between one overseas tour and another; in the same way the Victorians parked regiments in Ireland as a rest from India and at last set up the two-battalion regiments to make rotation easier.

In 1925, Inland and Coastal Areas were brought together under the command of Air Defences (*sic*) of Great Britain. This at least suggests that the units in the United Kingdom were here for some purpose. But the fighter squadrons were still stationed around London because there seemed no other function for them than defence of the capital, while the bomber squadrons were placed where they could conveniently co-operate with the Army and use its exercise areas on Salisbury Plain. There was no real change in this siting by 1931.

It is occasionally remarked that the RAF at this period was organized with France as a likely adversary. All this means is that at the time ranges were such that nobody but the French were near enough to fight, on one side of the water or the other.

The impression given by the *List* and the siting is that well into the thirties, the RAF in Britain had no independent war role, but saw itself as supplying the Army and the Navy with the air power they needed just as in 1916. For the rest, it supplied a presence in the Empire, which could conveniently be performed with what were alternately known as light bombers or general purpose aircraft, according to the equipment and armament fit. These aeroplanes could be used as required to overawe the dissident tribes by overflying and showing their capability of bombing villages, which they did when required; or they could reconnoitre for the Army, spot for the mountain batteries, take photographs for mapmaking, or carry the District Commissioner about his area more quickly than his horse or his truck could do. These tasks were done in the 1920s by Bristol Fighters or by the variants of the DH9 such as the Wapiti and the Wallace: in the 1930s they were carried out by variants of the Hawker Hart, such as the Hardy, the Audax or at the last the Hind. But they were hardly suitable weapons for taking part in a European war on the grand scale, and such a war was not at this time foreseen either by the Air Ministry or the War Office. Only the Admiralty, trying to make its case for buying and maintaining aircraft carriers, was obliged to consider the possibility of war with the United States or Japan, but not for another ten years. That was the rule.

Within this illogical organization, the squadron was seen, in folklore, as a fairly permanent structure. The squadron remained in one place for many years, regardless of the changes in command structure and in the aeroplanes with which it was armed. Once the unit had worked up, in personnel terms it did not so much change as flow, so that while there were always some newcomers among the pilots or among the airmen, at any given time the majority had been in post for a year or more. The squadron had a stable structure, which would enable it to withstand the strains of defeat, or of victory, and treat each impostor much the same.

It was this stable framework, of men doing what they were there to do, keeping their aeroplanes in action, automatically and without prompting, driven and kept efficient by loyalty to the unit, that Galland was thinking of and demanding. But by 1940, the old structure had vanished in the flood of dilution. Perhaps Galland would have been better advised to request an efficient group headquarters and staff officers, and a rational control and reporting system: because this was the new formation which was giving the RAF success against him.

Notes and References

1. For the Normal Tour of a Bomber Pilot, see W/Cdr. J. McLean, 'The RAF Training Year at Home', *RUSI* Journal, 1935, pp. 50 seq.
2. T. Heffernan. 'The Aircraft and Armament Experimental Establishment', *J.R. Aero Soc.*, 1966. G. Kinsey, *Martlesham Heath*, Terence Dalton, 1975, pp. 37-63.
3. For e.g., *His Majesty's Aeroplane, Carthusian*, see Owen Thetford, *Aircraft of the RAF since 1914*, Putnam, pp. 64, 262.
4. AP125, reprint of 1936, pp. 224-8.
5. Over several years, *Flight* devoted a good deal of space to articles on life in the cadre squadrons.
6. The Wapiti was not a bad aeroplane, except for its name: it was a very good aeroplane for the job it was meant to do in India. After all, it was a modified Wapiti which first flew over Everest, and that from a distant base. It was just kept in service long after it was quite outdated and thrown to the Auxiliary squadrons to wear out.

Chapter 6

Groundwork

What It Was For
Everyone knows Biggles. W.E. Johns, who invented him, had been a fighter pilot in the RFC, and then transferred to the RAF after the war. He was for a time the recruiting officer at Ealing, where he refused to attest T.E. Lawrence as a recruit to the RAF on the grounds that the man had at some time been flogged and was therefore medically unfit for service. That is another story. But Biggles first appears on a fine day in March 1918. The squadron is stood down, and though they do not realize it, but the reader does, or ought to, next day the final German offensive will open. Most of the pilots are sitting out in the sun on the edge of the airfield, but Biggles is down at the Flights. He is the old man of the squadron, he has been out eighteen months and is still alive, and this particular story is to show why. He is spending this rest day sitting on a kitchen chair turned round back to front, watching *his* fitter and *his* rigger at work on *his* aeroplane while he is filling his own Vickers gun belts himself, checking every round.

What Biggles knows well, and because he knows it he is still alive, is that men do not get killed because they are bad shots or because they cannot throw the aeroplane about. They get killed because in the middle of an action, a bulged round gets stuck in the extractor claw and the gun will no longer fire; or because in a dogfight their overstressed engines stop because the petrol is dirty and no one has actually seen it strained through a silk stocking, or because a valve which is superannuated or badly ground-in burns out; or because the strain on wrongly tensioned bracing wires breaks them and the wings just fold up. Men got killed because their aeroplanes were not properly serviced, and the only way for the pilot to be sure of that was to watch it done, and do himself the simplest things requiring no real skill.

Trenchard knew that everything depended in the end on the men who serviced the aeroplanes. 'We must use,' he wrote in his Memorandum, 'every endeavour to eliminate accidents both during training and subsequently. This end can only be secured by ensuring that the training of our mechanics in the multiplicity of trades . . . is as thorough as it can be made. The best way to do this is to enlist the bulk of our skilled ranks as boys and train them ourselves.' He was announcing Apprentice Training.

The Navy had begun to train engine-room artificers as apprentices in 1904. The Navy had always enlisted boy seamen, not so much for training as for service. The ERA apprentices were to be trained by the Navy and then to enter service as adult ratings, and not as boys. But the ERAs were always intended to be a small *corps d'élite*, and so they still consider themselves. They were not intended to be more than about one man in twenty. But Trenchard intended that the normal airman should be an ex-apprentice, joining the service at sixteen and entering on his man's service only after three years in the workshops and the classrooms.

The innovation went further than that in its implications. The Army had always recruited its other ranks from among the uneducated working class. Secondary education in any case had only become widely available with the Education Act of 1902. This empowered local education authorities to finance schools to give a secondary education, defined as a general education continuing to the age of sixteen. But this secondary education was not to be free. Fees were payable, although in many areas they were charged on a sliding scale regulated by parental income, so that a few poor boys were able to obtain a secondary education free. As a consequence, this secondary education in local authority schools was largely limited to the sons of the lower middle classes or of the upper levels of the working class, the sons of the foremen or of the skilled workers or of the corner shop. Although the poor lad could get his education free as long as he could stay at school, he might well be compelled by sheer poverty to leave school at the minimum legal age of fourteen and take a job, if he could get one. Even if he could not get a job, free schooling tends to be expensive. The schoolboy must be fed and clothed, and his fares paid, and there is a wide range of incidental expenses. Working-class parents of the time simply could not afford this kind of educational investment.

Entry to the aircraft apprentice scheme was to be limited to boys of sixteen years old at the time of enlistment, who should be nominated by their secondary schools or by a wide range of other organizations, such as the Boy Scouts. They had, in the first year, to pass an examination set by the Civil Service Commissioners, but almost immediately they were able to offer instead the new Schools (*sic*) Certificate.

The need for simple evidences of education had been seen early. The College of Preceptors had begun to set examinations for pupil teachers in 1852, and the Royal Society of Arts had offered certificates of proficiency in less academic subjects after examinations set at Mechanics' Institutes, from 1854. The Oxford Local Examinations Syndicate had begun to set its own examination for the convenience of private schools in 1856, with two levels, the junior at age fifteen or under, the senior at seventeen, or under. In 1917, the Board of Education set up the Secondary Schools Examination Council. This created the seven, later eight, Examining Boards which still exist, to set the examinations for the School Certificate at age sixteen and the Higher School Certificate at age eighteen.

The School Certificate was given after the candidate had passed, *at the same sitting*, in papers in five subjects, including always English, mathematics, a foreign language and a science. The details varied slightly from one board to another. The maximum number of subjects which the candidate could sit at once was eight, and therefore most secondary schools framed their timetables around the idea that every pupil would work for eight subjects.

It is not usually remembered that in those days the theory of the University of London (and of those universities which either were at first colleges which presented their students for external London degrees and which turned later into independent universities, or which were always independent but modelled themselves on London) was that most candidates would enter immediately on matriculation and take Pass degrees, after three years of study. A few would take Honours degrees, after an extra year of study. Matriculation was consequent on passing the School Certificate at a reasonable level. A candidate who could also show the new Higher School Certificate, taken usually in three subjects, could be allowed to take an Honours degree in three years instead of four. The change to the Honours degree as the mode, and to the Higher School Certificate as a condition of university entrance, was a gradual process through the 1930s. The School Certificate therefore should not be equated with the mere five 'O' levels usually demanded for apprentice or officer aircrew entry in the post-war period.

The requirement for the School Certificate was therefore a social statement that the bulk of the skilled tradesmen, and therefore the majority of the other ranks of the RAF, would be drawn from at least the lower middle classes, and certainly not from the lower working class like most of the Army recruits. It was also a statement that the bulk of the RAF entrants would be of an equal intellectual standard to their schoolfellows who could stay on and try for teachers' training college, or even for the university. More than that. School certificate was the normal entrance qualification for Cranwell or a short service commission.

The result of this selective policy was seen by T.E. Lawrence, in his service as a private in the Tank Corps. He compared the airmen with whom he had once shared a barrack-room with the soldiers he now knew on the same terms. The airmen had joined to advance themselves, while the soldiers saw their enlistment as the bottom step in a long social descent. 'The fellows at Uxbridge had joined the RAF as a profession – or to continue in it at their trades ... These fellows (soldiers) have joined up as a last resort, because they have failed and were not qualified for anything else.' The airmen, on the other hand, saw their service as the beginning of a real career. They had as little to do with their officers as possible, and this, says Lawrence, was a mutual avoidance. The men, to Lawrence's observation, saw themselves as the equals of their officers, at least in competence. ' . . . the (RAF) officers were treated by the men off parade as rather humorous things to have to show respect to'. Lawrence, who was not the most perceptive of observers in any social setting, could not understand why, but his reader, who in this case was an air vice-marshal, could see the point.[1] This very clear distinction between the services was obvious to the men in the ranks, and led to the term 'Brylcreem boys', a nickname which mixed derision with envy.

As a result of the necessities of flying, Trenchard was saved from the obloquy which has attached to the generals of the period between the wars. They never quite learned the demands of servicing a machine, and consequently they accepted the old definition of a machine as a thing that goes wrong. Their introduction to the fighting machine came with the tank. The first tanks were tried on ground cut up by shelling and flooded, so they failed. But in 1917, a great set-piece attack was mounted at Cambrai, and can only be compared with the first thousand-bomber raid. Here the training battalions were stripped so that somehow 476 tanks could be gathered for a great assault, over firm hard ground, unshelled, undamaged. Only sixty-five were actually knocked out by enemy fire, but by the end of the first day forty-three were irretrievably bogged down and seventy-one had cast their tracks, had burnt out their engines, or broken down for a variety of other reasons: the surviving crews were just too exhausted to fight next day.[2] The generals' experience of aeroplanes was almost as bad. But horses, well, they knew horses. Horses would carry on working at least as long as the men would; dirty, half-starved, mangy, fleabitten, even wounded, they would carry on to the end. So the soldiers wanted to keep horses, not for the great romantic charge, but to pull the guns and the ammunition wagons and the ration carts. And at the very worst, you could always eat the horses.

The Apprentices

Trenchard, then, put in motion the training of apprentices by the RAF. There were to be three trades. The larger trades, the fitters or engine workers and the riggers or airframe workers, would be trained at Halton. The smaller group, the wireless trade, would be trained at Cranwell as a lodger unit on the RAF College. The system was originally announced as the Boy Mechanic Scheme in November 1919, but had officially become the Apprentice Scheme by the middle of the following year.

The formal announcement came on 26 April 1920. Candidates

were to be nominated by Local Education Authorities, or by the governors of certain named schools. There seems to have been a feeling at this stage that apprenticeships would be attractive to pupils leaving public (that is fee-paying boarding) schools. The candidates would be medically examined after they had passed the competitive examination, set by the Civil Service Commissioners at fifteen local centres, and covering mathematics, experimental science, English and a general paper. There were to be 300 vacancies for the first entry. Apprentices would be attested for ten years regular and two years in the reserve. Outfit, lodgings and food would be provided free. Apprentices under eighteen would be paid one and sixpence a day, three shillings a day for the over eighteens. Those who wished could buy themselves out for £20 during their first three months, or for £100 thereafter. The first draft of boys left London for Cranwell on 18 January 1921.

The conditions of service life at No. 1 School, Halton, were described in an article in *Flight* in October 1923. The apprentices worked in the classrooms and the workshops from Monday to Friday. Wednesday afternoon was a sports afternoon. Report cards were made out at three-weekly intervals. Saturday morning was given over to drill and inspections of boys and huts. On Sunday morning there was church parade. Apprentices were allowed out of camp on Wednesday afternoons after sports, and on Saturday and Sunday afternoons. Those over eighteen were allowed to smoke, but only out of camp. The word camp was already being used colloquially for the Station.

We know nothing of the everyday lives of these first apprentices. Later evidence encourages us to believe that like those who followed them, these first trainees developed a complex social pattern among themselves, with hierarchies of seniority and of trades, traditions of who got served first in the canteen, of who were entitled to sit near the counter there, or who could claim the best bed-sites in the huts.

In November 1924, the first passing out parade was held at Halton, and the press were sparsely in attendance. The original intention was that the course should last three years, after which the bulk of the entry should pass out into adult service as leading aircraftmen, a few as AC1s, but some even as corporals. The importance of the occasion was to be marked by the attendance of the Chief of the Air Staff himself, who would announce, it was thought, his satisfaction with the results. But little did the assembled company know Trenchard. In a storming speech he slated the apprentices and by implication all those who had taught them. He was, he said, bitterly disappointed with the results. The number of LACs and AC1s was too small. He did not mention corporals at all. After a few more well chosen words, he directed that the AC2s and those who had actually failed the course should be kept at Halton for a further six months and pass out with the next entry.[3] The press were somewhat taken aback; for the most part they had never seen Trenchard in wrath. The following week the first Cranwell entry of wireless tradesmen passed out, and Trenchard was there too. This time the press attended in overwhelming numbers, wondering what he would say now. But the Chief of Air Staff was pleased with the results here, and the reporters were disappointed.

These boys were being trained for a Royal Air Force which was essentially the same as that of the war. Most aeroplanes were built of wood, and occasionally the riggers are referred to as carpenters. They were required to be able to repair or remake any part of the aeroplane, and it was stated during the first years of the scheme that each entry of fitters and riggers would build aircraft as part of their training; the first entry would built three Avros, the next entry three Grebes and so on. It would be interesting to know if the aeroplanes were actually built and who, if anyone, was brave enough to offer to fly them.[4]

The innovation of an educated body of non-commissioned officers and airmen was in itself a social revolution. It was very much a favourite project of Trenchard, and it is unlikely that he failed to understand the implications. Equally revolutionary was the proviso that of each entry, the best three apprentices should be awarded cadetships at the RAF College at Cranwell, which was to prepare them together with the normal entrants for full careers in the service, with promotion possible, and even likely, to group captain and beyond.

The Army had never thought of anything like this. The occasional outstanding ranker, the Wullie Robertsons, had been commissioned with some difficulty. During the First World War, men in some numbers had been commissioned from the ranks, and an Officer Training School had been set up in France in

1915. Robertson, in his autobiography, says this was his idea. During the years from 1906 to 1912, the Army had set up a scheme whereby NCOs who had passed the relevant Army education tests could claim as of right to sit the Promotion Examinations from second lieutenant to lieutenant and from lieutenant to captain. As a result, when the war started in 1914, three quarters of the sergeants in the army were qualified for captaincies, and were commissioned. But nobody had ever suggested that the ranks of the recruits to the Army should be combed for likely officers to be sent to Sandhurst. Sandhurst was for gentlemen who could afford to pay the fees.

So the very best of the apprentices could hope for early promotion to the officer ranks. But there were other ways up in the world, for apprentices and for other airmen. Both types of airman could hope for immediate promotion to sergeant if they were willing to pay what Trenchard, believing aviation to be war and not a sport, considered to be a high price. They would have to learn to fly.

One of the most perceptive novelists to describe an air force at war has remarked that 'unless you start to fly when you are very young, you are unlikely to feel complete confidence in the strength or ability of any aeroplane'.[5] Trenchard had learnt to fly at an advanced age, and could not really conceive of it as anything more than a hazardous experience. It would seem that he had been more impressed by the casualties in the RFC than by their achievements. Everybody in the years between the wars lived conscious of the dreadful slaughter of the Somme and Passchendaele, and were determined, generals as well as pacifists, that this should not occur again. It explains the continuing stress in the Air Plans between the wars on strategic bombing, giving grounds for the hope that a new war could be won by bombers with small losses. Any losses would be among the pilots. Learning to fly, in Trenchard's imagining, was a burden imposed on the officers, but the men who had volunteered to go to Calabar to wait in a pool for casualty vacancies would be willing to learn to fly. The trouble was, there might not be enough of them.

The announcement came in autumn 1921 of the establishment of a class of airman pilots. Qualifications included 'a high standard of education and efficiency'. Candidates would also have to show that they possessed the qualities of 'pluck, reliability, alertness, steadiness, keenness and energy'. Preference would be given to men already qualified as air gunners, that is to men who knew what flying was about. They must serve for five years from the start of training: men would not be recruited with the promise of pilot training. Men, whether apprentices or other tradesmen, would at the end of their tour of duty of flying return to their trades but with the rank they had gained in the air, that is with accelerated promotion.[6] The net effect of the scheme was all that was hoped. By the early 1930s and until the war started, about a quarter of all RAF pilots in squadrons were NCOs.

What were the trades to which the non-apprentices would return? These were men who were sent by the local recruiting offices to Uxbridge, to be trained as recruits. The training here was much the same as it was for any recruit to the Army. A fair amount of parade ground drill had to be learned by any large body of men who were going to live in big concentrations in camps, or on the stations. Simply getting them around the place, between barrack rooms and messes for their meals and the classrooms and workshops for their training needed a fair amount of organization. But the men, Lawrence tells us, were 'squeamishly tired of being thought to resemble the Army'. The general public were still in the early days not prepared to accept that the new RAF was a military force. *Flight* remarks in surprise at the appearance of the RAF in BLUE (!) at the Royal Review of 27 April 1920 with rifles and fixed bayonets. Three years later, *The Times* correspondent at the first passing-out parade at Halton notes that there were 1,300 boys on parade and then says that it was 'strange to see bayonets in conjunction with the blue uniform of the Air Force'. Yet any examination shows beyond doubt that the RAF was organized and worked exactly as if it were an Army organization, one of the Scientific Corps.

In *The Mint*, T.E. Lawrence describes the experiences of an entrant as an aircraft hand general duties, that is the completely unskilled man. The first two sections of the book merely describe the basic military training which any recruit would get at a regimental depot; except for the sight in print of the foul language which it was not at that time usual to put down on paper, the book would not have surprised anyone who had done his military service in the First World War. We must remember that although Lawrence had been a colonel he had never been a private, and until he first joined the RAF had no contact with members of the working class except in giving orders or in the way of trade.

It is only Part III of *The Mint* which has any specific RAF content; here it describes the easier disciplinary conditions of a force in permanent garrison in the home country, and also the common link between all ranks, a fascination with control of the machine. Status in this service was accorded to the man who controlled the machine, either the aeroplane or the Brough Imperial, the motor cycle which seems to have given Lawrence a standing in the service community above his rank. And Lawrence describes with sympathy the differences between his unskilled equals working on the apron at Cranwell and the newly entered Wireless Apprentices.[7]

After two months at Uxbridge the recruits went to The School (later No. 3 School) of Technical Training (Men) at Manston. *The Times* had in 1919 already announced the opening of a new technical school near Ramsgate. This had 1,000 men in training: Number 1 Section took drivers MT, fitter-drivers, fabric workers, vulcanizers, and general fitters. Number 2 Section took riggers and carpenters to cover the gap before the Halton output began to come in. Men with civilian qualifications in their trades would not be sent to Manston (Table 19).

In July 1925 it was announced that the RAF would now recruit boy clerks. These would be recruited partly by competitive examination (thirty per year), and partly by direct enlistment of fifteen per year with School Certificate. They would be trained for two years in shorthand, typing, bookkeeping and practical office work. No candidate could compete at the examination for clerks and at the same time for an apprenticeship.

Aeroplanes
We must remember that in the very small Air Ministry, Trenchard still kept a rigid control on the design and purchase of aeroplanes. True, he could not make any massive purchases with the money at his disposal, and he had to keep a close hand on the purse strings. It is usually said that Trenchard was tempted in 1925 to an airfield to watch a display by the new Fairey Fox day

bomber, a private venture, and that he immediately ordered a squadron of the new aeroplanes. One squadron would represent a large proportion of the bomber force. What is important about this story is not whether it happened; it is that everybody in the aeroplane business thought that Trenchard was perfectly capable of making such an impulse buy and putting it through. He seems to have had the authority to do this. But there was planning. A squadron of Foxes meant one squadron less of some other type.

The Fox story is typical of aeroplane buying through the interwar period. Again and again the RAF buys, or is put under pressure to buy, a light or day bomber which outperforms the fighters of the time. The Fox outdid the Siskin, the Hart surpassed the Bulldog, the Blenheim was faster than the Gladiator, and so was the Battle, of which *Flight* said 'at last a real aeroplane for the RAF!'

There were Air Ministry specifications, and these were not very stable things, since sometimes they were issued before aeroplanes were ordered or even built, sometimes they seem to have been made up to fit the aeroplanes being offered. Whichever way, the specifications were in the Trenchard era very detailed things. They laid down, for example, the location of the petrol tanks, always outside the fuselage, the routes of at least three different fuel feed-lines, the armament and the amount of ammunition to be carried, the cruising speeds at set heights, and the landing speeds, usually for a big twin-engined bomber as low as 55 mph.[8]

In fact the specs carry the imprint of the earlier Trenchard who went around the squadrons, listening to his pilots and working later through the visit reports he had made. The aeroplanes of the late 1920s look like the aeroplanes of the First World War because they embody all that war's first-hand experience as the pilots gave it to Trenchard.

The aeroplanes were all biplanes, because biplanes worked. The box-kite structure provided the strength required. The trouble with monoplanes in those days of wood was that the main spars could not at the same time be built both strong enough and light enough to take the strains required. The monoplane as late as 1930 was a clumsy thing, held together by little masts and nests of bracing wires, and by struts. The first real monoplane in the service, smooth and sleek with its cantilever main spar and undercarriage wound up by hand, was the Anson. It was the first monoplane which did not break up when you flew it to the limit of its engine performance, which of course is what you have to do with a military aeroplane.

The same thing applied to fighter armament. As early as 1922 it became obvious that a fighter armament of two forward firing guns did not give the rate of fire needed for a quick kill. But how to put more than two guns on a wooden biplane with limited engine power? Each extra gun meant at least half a hundredweight extra load in itself. A synchronizing gear could be fitted to an engine to cope with the fire of one gun, or two: but to synchronize more guns meant a great deal of machinery, and every extra complication in the gear meant a greater likelihood of breakdown. There was an attempt to fit the Hawker Demon, a two-seater fighter with metal framework and fabric covering, with three forward-firing guns on the fuselage, but it cannot have been very successful, because it was not continued. The first four-gun fighter was the Gladiator, which carried the conventional two synchronized guns on the fuselage but had two more guns inside the lower wings, outboard of the propeller arc and therefore not needing synchronization. The wings of the earlier monoplanes and biplanes were simply not strong enough to take the weight of the gun-mountings or the strain of firing.

And the guns in the Gladiator were different. Originally the gun used in fighters was the Vickers, and the flexible gunners in the bombers or the two-seater fighters had the Lewis. Most of the training of the Army Vickers or Lewis gunner was spent on learning the various kinds of jamming the gun is liable to, and the corrective actions. If the pilot could reach the guns then at least he could do something to clear jams when they happened: a gun buried in the wing would be lost to any attention. The Vickers would not do for multi-gun fighters. In 1934, the Air Ministry invited competitive tenders for new guns, which would be buried in wings out of reach of a pilot, and therefore beyond any hope of correction of jams: reliability would be the main requirement. The trials were held at, of course, Martlesham Heath. The winner was the American Colt company, and their gun, renamed the Browning and manufactured by a number of British subcontractors, became the standard armament.[9]

The normal weapon of the flexible gunner in the bomber, the

Lewis gun, was also replaced. Ammunition supply for this gun was limited. The feed was from flat disc drums, which in the original ground design held thirty-three rounds, and jammed easily: there were triple layered drums for the airborne version, giving ninety-nine rounds before the drum had to be changed, and cubing the chances of jamming. The task of the gunner, standing up in the slipstream, trying to change a drum or clear a jam with heavily gloved hands, became quite impossible with increasing speeds. The enclosed power-driven turret came in with the Overstrand in 1934. All the arguments for multi-gun fighters applied to the defensive gun-positions in bombers. The new gun was the Vickers K, belt-fed, and mounted, once the engineers got courage, in two or four gun clusters in turrets.

The old guns and the old aeroplanes stayed while Trenchard was in the driving seat. In 1931, C.R.F. Fairey remarked that up to that date the specifications issued by the Ministry had always been very detailed and rigidly insisted on. That had ended. Now the specifications were issued in general terms, and would suggest, for example, that 'the special requirements set out in para 7 (dealing with the accommodation of crew and armaments in the fuselage) are to be regarded as ideal. They need not be rigidly adhered to at the expense of performance.' The attitude of the service had changed. He could not think why.[10] But with Trenchard's passing, many things had changed. Above all, the fighter pilot would now be dependent entirely for his life on the apprentice-trained armourers who loaded his belts and serviced his guns, the ex-apprentice instrument makers who set up his panel for flying blind in cloud and at night, the Cranwell-trained wireless fitters who kept his radio serviceable and saw that the radar stations kept on supplying his information, and the ex-Halton fitters who kept his Merlins in tune.

Notes and References

1. T.E. Lawrence, *Letters*, ed. David Garnett, Jonathan Cape, 1938. Note particularly letter of 9th November to Air Vice-Marshal Swann: note also letters to other friends of 27 March 1923, 14 May 1923, 13 June 1923, 19 May 1925, 2 December 1927, 15 January 1928.
2. Liddel Hart, *The Tanks*, Cassell, 1955, p. 135 et seq. His figures

are confused but seem to add up to only fifty runners surviving for the second attack.

3. *Flight*, 25 November 1924.

4. *Flight*, 4 October 1923. Wing Commander E.D. Breese, *Journal of the Royal Aeronautical Society*, vol. XXXI, 1927, p. 219. Note that in the paper and in the discussion, there is already criticism of the teaching of too much drill to apprentices, and a call for the establishment of an Engineer Branch in the RAF.

5. James Gould Cozzens, *Guard of Honour*, Longman Green and Company, 1949, p. 77.

6. *Flight*, 15 September 1921.

7. T.E. Lawrence, *The Mint*, Jonathan Cape, 1955. While he passed as Shaw and Ross, Lawrence's real identity was well known to the airmen he served with: they regarded him as that not unusual figure, the gentleman ranker, whose anonymity was to be preserved against the Press and other outsiders. Personal communication, Mr V.F. Strickland.

8. Compare the specification of the Heyford, Spec 19/27, following the old pattern, in C.H. Barnes, *Handley Page Aircraft since 1907*, Putnam, 1976, p. 288. The Heyford certainly did not look like any aeroplane of the First World War, or indeed like anything else at all.

9. G.F. Wallace, *The Guns of the Royal Air Force*, William Kimber, 1972. Chapter 5 traces the development of the Browning gun from a buy of six sample guns in 1926. Chapter 6 describes the development of the Vickers gas-operated gun or K gun, from its origin in the Vickers-Berthier gun of 1919. For the supply of the weapons, see Donovan M. Ward, *The Other Battle*, published by B.S.A. in 1946. The first delivery date for the Browning was September 1936.

10. C.R. Fairey. 'The Future of Aeroplane Design', *RUSI Journal*, 1931, p. 563. This was a paper read to an audience on 11 February 1931. Air Vice-Marshal Dowding was in the chair. The discussion included pressure for more gunpower in fighters.

Chapter 7

Worse Things Happen at Sea

A Sea Change
It is common knowledge that one of the features of the interwar years was the dispute between the Air Ministry and the Admiralty about control of the pilots and aeroplanes embarked in the carriers. It began with face to face arguments between Trenchard and Beatty,[1] and ended in 1937 with the formation of the Fleet Air Arm. Was this any more than a storm in a teacup?

The Navy came to the great date of 1 April 1918 with a naval air arm called The Royal Naval Air Service. With splendid impartiality, the *Army List* of 1914 lists all the naval stations and officers, as the *Navy List* shows the Army units and officers. At the outbreak of the war, the RNAS had a Central Air Office, the Naval Flying School at Eastchurch, with Samson as the Officer Commanding, and naval air stations at Isle of Grain, Calshot, Felixstowe, Yarmouth, Fort George and Dundee. There was an airship station at Farnborough, with Naval Airships Nos. 3 and 4 (Tables 1 and 3).

In 1918, all the naval stations and squadrons were handed over to become integral parts of the new Royal Air Force. It is said that the RFC pilots resented the new order, because they had been happy as part of the Army; on the other hand, the naval pilots welcomed the change, because the Navy, they thought, had never really appreciated them.[2] The usual phrase attributed to Beatty was that the Navy wanted sailors who could fly, not pilots who were in the Navy. But the Navy's policy seems to have been directed simply to avoid attaining that. In contrast the Army had always got what it wanted, soldiers who could fly. The naval aviators as a group felt that they had wished themselves on the Navy which wasn't really sure if it wanted them. Some sailors wanted them. Jellicoe, for example, in 1916 was writing that a Zeppelin was worth a good many light cruisers.[3] For want of air reconnaissance, when the High Seas Fleet came out twice again in 1916, after Jutland, the Grand Fleet was unable to catch it. The Grand Fleet was depending on a submarine screen for its information, and so were the Germans, in addition to the Zeppelins. On each occasion, the submarines of each side did a great deal of damage, sinking a British cruiser, and putting three German battleships into dry dock.[4]

To deal with their reconnaissance problems, and even to carry out offensive operations, the Navy began to experiment early in the war with methods of taking aeroplanes to sea. The first attempts entailed taking seaplanes out on carriers. Later they went on to taking land-planes, and having them come down in the water alongside their carriers on the return, as if they were seaplanes. Seaplanes could be picked up again but their performance was inferior: the land-planes, whose performance once in the air was much better, were simply abandoned when the crews were recovered. The need was for ships on which wheeled aeroplanes could land as well as take off. The Royal Navy was the first in the world to try building landing decks on the tops of hulls (Table 11).

As early as 1914, the Admiralty had begun to convert a cargo ship for use as a seaplane carrier. This ship was named HMS *Ark Royal*, and carried five float planes and two land-planes. She served in the Mediterranean as an aircraft depot ship till the end of the Second World War, although she was renamed *Pegasus* in 1935 to release her name for a new aircraft carrier. There was also another converted merchant ship bought on the stocks and fitted out to carry nine seaplanes, under the name of HMS *Pegasus*. She remained with the Mediterranean Fleet as a seaplane carrier till 1923, and was then rated as an aircraft tender till she was sold for breaking up in 1931. During the 1930s the idea of a seaplane carrier died out.[5]

Building carriers in a hurry depended on having hulls available. The first experiment used the hull of a liner being built for an Italian line, and left lying on the stocks since August 1914. This hull was fitted with a complete flush deck, and was successful both as a launching and as a flying-on ship. The complement, or theoretical capacity, was twenty aircraft. This ship, HMS *Argus*, was still in use at the end of the Second World War. The next successful ship used the hull building on the Tyne for a Chilean battleship and also abandoned at the beginning of the war. This gave a ship with an island superstructure to starboard, just fast enough to keep up with the battle fleet. She became HMS *Eagle*, which had a complement of twenty-one aircraft. She was torpedoed in the Mediterranean in 1942. The first purpose-built carrier for the British Navy was HMS *Hermes*, about half the size of *Eagle*, but also with a complement of twenty aircraft: she was sunk by the Japanese in April 1942.

During the Second World War it became clear that British carriers were able to operate on average about 25 per cent more aeroplanes than their official complement: nevertheless nobody knew that till they had to try it and it was the complement figures which were used for planning. The Japanese, who had their first purpose-built carrier ready before *Hermes*, usually were only able to operate about 25 per cent fewer aeroplanes than their published complements. The French built a carrier, *Bearn*, which was meant to carry forty aircraft but was so small and cramped that no more than ten of the aircraft carried in the hangar below could be brought on deck ('ranged') and operated at a time. All nations learnt that there were limits to carrier size. If the carrier were too small, it pitched too much to be usable in moderate weather: if it were very big, the number of aircraft which could be operated was still limited by air traffic problems.

After 1914, the Navy still clung to the dream of a landing in the Baltic. Three ships were built for this task, and known as 'light battle-cruisers'. They were each of about 23,000 tons, the same size as *Eagle*, and fitted with a small number of very big guns. As the scheme was impracticable and there was no other use for them, the experiment was tried of fitting one, HMS *Furious*, with a flying-off deck. This worked, but there was no really easy way of landing on the temporary deck as it was then built, with a funnel sticking up in the middle and generating hot fumes and turbulence. The three hulls were so attractive that the Admirals could not bring themselves to scrap them, and decided to turn them all into aircraft carriers with full-length flying decks. Each carrier would provide a post for a senior captain and three or more commanders. The big guns were taken out: the fifteen-inch guns and turrets for HMS *Courageous* and HMS *Glorious* were kept and used twenty years later to arm the new battleship HMS *Vanguard*: the two eighteen-inch guns from *Furious* were first put into a monitor, and later mounted at Singapore.

Finishing the conversions took some time. *Furious* was ready in August 1925, *Courageous* in April 1928, and *Glorious* in March 1930. The next carrier built, designed from the start as a carrier, was HMS *Ark Royal*, with a complement of sixty aircraft: she was ready in November 1938, after the formation of the Fleet Air Arm. These four big carriers could make thirty knots and could therefore operate with a battle fleet or with a cruiser force, while *Argus* had a speed of only twenty knots, *Eagle* was rated at twenty-two-and-a-half knots and *Hermes* at twenty-five knots. The Admiralty's problem was therefore how to find crews for the thirty-six aircraft in *Furious*, and the forty-eight each in *Courageous* and *Glorious*, as well as for the hundred or so in the three small carriers, that is about two hundred and thirty in all.

The time was not important. On 5 August 1919, the Admiralty suggested the convenience of assuming freedom from war with a Great Power or combination of Great Powers for a set time in the future, to allow them to plan their post-war rebuilding rationally. In view of the current Japanese building programme, they suggested that five years would be convenient. The Cabinet Committee on Finance, consisting on this occasion on 11 August 1919 of Bonar Law, Milner and Austen Chamberlain, made this suggestion the basis of a general rule, and set the period at ten years. The rule was renewed from year to year, and in July 1928, it was fixed on a moving day-to-day basis. This rule is usually blamed for slowness in re-armament. Other political events during the late 1920s probably had greater effect. The rule was quietly dropped in 1932.[6]

We must remember that the Admiralty in 1918 had to start organizing their air arm virtually from scratch. There was very little experience of operating aeroplanes from a floating deck. They had already built or converted three ships and committed themselves to the expense of converting three more. Now manning them was taken out of Their Lordships' hands, and given to another service who would provide, or at least had promised to provide, lodger units to sit in their ships and carry on whatever war their masters in Kingsway might direct.

In Practice

Carriers proved to be very expensive to build: they were ferociously expensive to operate. Whenever a carrier wished to fly off her aircraft, or land them on, it was necessary to turn her into wind and work up to as near full speed as could be managed in the time. Operating engines at full speed tends to wear them out, and also to put a strain on the hull itself. Normal warships were at this time being ordered to avoid high speed manoeuvres not only to save on fuel but also to escape damage to the ships.[7] But the carriers could not be operated at all without these high speed manoeuvres, and so tended to wear out very quickly. As a result, at any given time it was likely that at least one of the small twenty-aeroplane carriers and, when they appeared, one of the large carriers would be in dock for a refit.

Practical experience in flying off and landing on carriers was necessary for the individual pilots. Slowly, the limits of weather were explored, and carrier operations ruled as feasible in worse conditions than anyone had thought possible. It was even worked out how to take off and land on a carrier at sea at night. But all this wore out the ships, and brought casualties both in pilots and in aeroplanes.

These exercises required a great deal of staff planning, and the staff could not always be guaranteed to be aviators or to have any real understanding of the problem. Skill in handling carriers in action could only be expected of men who had themselves been pilots, and who were also ship-handlers. Two examples to illustrate this come from the Battle of Midway in 1942. As the two fleets began to make contact, a Japanese scout plane returned with an unserviceable radio and the information that at its limit of range it had glimpsed fleetingly an American carrier heading east, that is away from the Japanese carrier force. The Japanese Admiral Nagumo, a non-aviator, interpreted this as a retirement by the Americans. Only half an hour later did the aviators lay out the weather information on the plot and realize that what had been seen was a carrier turning into wind to fly off her aircraft: at that moment the aircraft from *Enterprise* and *Hornet* began to arrive over an unprepared enemy.

These US carriers were commanded by a non-pilot admiral, Spruance, with no previous experience of carriers. When the information as to the Japanese fleet's position arrived, the air commander in the carriers gave what was then a traditional order in the US Navy, 'Pilots to your planes!' The admiral asked how long it would be before the range was close enough to launch his aeroplanes, and was told that it would be two hours at least. Spruance then ordered the pilots stood down, and when they did take off to attack they were that much fresher. He thus earned a reputation among the pilots of being an indecisive cruiser admiral who did not know his own mind. There is no justice. But he won his battle.

Building up a body of experienced aviators who were also ship-handlers and who could therefore both handle carrier fleets and be taken seriously as leaders by the pilots was going to take time. The Admiralty had no hope of doing that if the greater part of their aviators were going to be men who came to fly at sea as only one episode in a varied career. But in justice their men could not be deprived of their chance of qualifying for big ship command.

The agreement with the RAF was simple. The RAF, in Trenchard's vision, was to provide the training of all pilots, and all the relevant services. The Navy was to allot a number of officers each year to train as pilots, up to a maximum of 70 per cent of the pilot vacancies. The problem was that the Navy could never find enough of its officers willing to train as pilots and spend one commission in the Air Arm. Even a scheme to take officers into the Air Arm for fourteen years, with an inserted two-year break part way through to serve as sea officers and get themselves up to date, was felt to be unsatisfactory. There were recurrent disputes, in which again and again the RAF offered to supply pilots. When it seemed particularly appropriate to pour oil on troubled flames, the RAF sportingly offered to supply enough airmen pilots to fill the vacancies.[8]

The reasons why naval officers did not volunteer in sufficient numbers to fly were also simple. The way to promotion in the Navy, to the command of a big ship or of a fleet, was to follow a varied career in what is now called the seaman branch, the mature officer alternating command of small ships with being first lieutenant in larger ships. A commission spent in the air, let alone a longer period, would come as an interruption in this career and would give no relevant experience in command or in ship handling. Flying was therefore seen as a dead end job. There also seems to have been a feeling in the fleet that an officer who left his 'natural' speciality, either as a sailor or as a Marine, to fly was in some undefined way letting the side down and implicitly criticizing his comrades and the established order.[9]

The Facts
It is interesting to see how many pilots were actually put into service with the Fleet, and how they were divided between RAF and Navy officers. The system worked by carrying, as part of the permanent complement of a carrier, a miniature RAF station. The senior officer was usually a squadron leader. He had under him three flight lieutenants, not there simply because they were pilots (which they were) but because they, like him, showed after their names in the *List* the symbols 'e' or 's' or 'p' or 'a'. That is, they had completed the in-service engineering, or signals, or photographic, or armament officers' courses. There was also usually an officer of the Stores Branch. When the aircraft and pilots were not at sea they were held at land stations, but the 'station headquarters' and its officers remained aboard. The Navy needed time to produce officers to fill these station headquarters posts.

In 1923, *Ark Royal* and *Pegasus* were in the Mediterranean, and between them had twenty-six pilots. *Argus* was in home waters and came under Coastal Area: she is listed as having a wing commander RAF and four flying officers, and was presumably waiting for her aircraft to fly on. Even if she were made up to full complement she would not have needed many more than two dozen pilots: the Gosport base was shown in the *List* as having a wing commander, a squadron leader, and twenty-one junior officers, which we may assume to be the missing crews waiting to fly on to *Argus*.

The actual requirement for the Fleet would therefore be around fifty pilots. At that time there were 694 pilots listed in the RAF squadrons, so the naval requirement was then a very minor thing. The air marshals and the admirals were arguing over the principles. In terms of numbers of pilots, the naval commitment was always a minor detail for the RAF: the cost of the carrier force meant that it was always of vital importance to the Navy. What was important to the RAF was the number of senior officer posts available to make careers for officers, not merely afloat but in the various headquarters staffs and in the permanent complement of one extra flying training school.

The third big carrier, *Glorious*, came into service in the spring of 1930, and we may assume that she had finished with working up and all the teething ailments that ships are heiresses to by the *List* of May 1931. There are no entries in this year for *Furious*. It may be assumed that she was now refitting, with *Glorious* to fill the gap.

Of the four carriers now active, we find that *Eagle* has a headquarters unit of a squadron leader and three flight lieutenants, one each annotated as 'e', 'a' and 's', with one Stores Branch officer. She carried three Flights, 402 Fighter, 448 Spotter-Reconnaissance and 460 Torpedo. They had listed by name six RAF officers and twelve naval officers. *Hermes* has an HQ unit with a squadron leader, flight lieutenants 'a' and 'e', and a Stores Branch officer. She carried 403F and 440SR. They have listed five RAF officers and nine naval officers. *Courageous* has two squadron leaders and four flight lieutenants, 'a', 's' and 'e', and a Stores Branch officer in the headquarters. She has 401F, 404F, 407F, 445SR, 446SR, 449SR, 450SR, 463T and 464T, with a total of twelve RAF officers and twenty-four naval officers. *Glorious* was now in the Mediterranean. Her headquarters had two squadron leaders, both 'e', three flight lieutenants 's' and 'a', and a Stores Branch officer. She carried 405F, 406F, 408F, 441SR, 447SR, 461T and 462T. These add up to thirteen RAF officers and twenty-six naval officers. Ashore between Gosport and Lee-on-Solent we find 442SR, 443SR, 444SR, 475T and 466T, presumably the units from *Furious*. They have ten RAF officers and fourteen naval officers.

That is a total of forty-six RAF officers and eighty-five naval officers. This comes close to the agreed balance of 70 per cent naval pilots in the carrier units. It also supplies almost enough pilots to fill the aeroplane complements of the carriers in commission. The unit, be it noted, is the Flight, with a strength of six pilots. The number of officers in the station HQ does not vary with the size of the ship.

The *List* of May 1936 was issued on the eve of the Admiralty's opening a new campaign for the recovery of at least the squadrons with the fleet, and if at all possible of the reconnaissance squadrons and bases which were to form the bulk of Coastal Command. We find that *Eagle* and *Argus* are not listed, and therefore were either being refitted or laid up. The unit now being used was the squadron, and these all had numbers in the 800 series. *Hermes* was in the Far East, and is listed as having two squadrons, 803F and 824 Spotter-Reconnaissance. But these two squadrons between them have listed by name only thirteen pilots, of whom five are naval officers and the rest RAF.

Furious was then in home waters, and had three squadrons: 801F, 811 Torpedo Bomber, and 822SR: the lists show by name thirty-four officers in these squadrons, of whom eighteen are naval officers, beside one airman pilot. *Courageous* was also in home waters, and had 800F, 810TB, 820SR and 821SR. These squadrons show twenty-five RAF officers, twenty-five naval officers and five airman pilots. *Glorious* was in the Mediterranean, and shows 802F, 812TB, 823FSR and 825FSR. Between them these squadrons listed twenty-four RAF officers and twenty-seven naval officers.

The carriers therefore accounted for seventy-two naval officers, eighty-one RAF officers and six airman pilots. The manning situation, from the Navy's point of view, had deteriorated. At this time, the lists show that the land-based RAF squadrons had in post 903 officer pilots and 263 airman pilots, a total of 1,166 pilots employed. Under training there were twenty-three naval officers at 1 Flying Training School: against this we find 345 RAF officers under training at the other nine Flying Training Schools in the UK and 4FTS in Egypt, although there is no record here of the number of airman pilots under training. There was, of course, no question of expanding the Fleet squadrons till the new carrier, *Ark Royal*, should be completed in late 1938, but it does not look as if there were any hope of manning her entirely with naval pilots immediately even then.

After the formation of the new Fleet Air Arm, with little time left before the threatening war, the Admiralty began to offer short service commissions for pilots in the RAF manner. These were not attractive to enough applicants. The Navy still clung to the principle enunciated by Mr Pepys in the seventeenth century, that the care of a King's ship was only to be entrusted to a captain or, as a watchkeeper, to a lieutenant (the word means 'deputy') who could show that he had satisfied his seniors as to his competence and had also spent a sufficient time at sea in a lower rank. This was also applied in a higher degree to admirals: anybody could be a general, which was a political appointment, but the safety of the realm was in the hands of the Royal Navy and only experience and competence would do. A man engaged only to fly would not have either this competence or this experience. The ambitious young man whom the Admiralty hoped to attract could see that if he took a short service commission in the RAF he had at least a sporting chance of getting a permanent commission and eventually commanding a squadron, or a station: while if he went into the Navy as a pilot, there was no chance that he would ever command anything at all.

It was obvious then to anyone who had access to the *Lists*, that is all the foreign air or military Attachés who get them issued free, and any stray civilian with four and sixpence in his pocket, that although the four carriers in commission had a complement of 152 aeroplanes between them, there was some difficulty in providing pilots; the three large carriers were just barely manned with no reserves on the squadron strengths, and that had only been managed by stripping the carrier in the Far East. No speedy reinforcement would be possible; to borrow more pilots from the RAF would mean tying up one carrier for some time on the deck landing training task, quite apart from conversion time to learn the art of flying the naval machines and carrying out the naval task. To fill the vacant aeroplane seats from pilots now returned to normal naval posts would mean giving them refresher flying and working up time, as well as leaving large holes in the Fleet organization. Even the desperate expedient, adopted in 1938, of training rating pilots would not give enough men in time.

And in spring 1936 speedy mobilization and deployment of the Fleet was definitely on the cards. The previous autumn, Italy had gone to war, and invaded Abyssinia. A Fleet concentration at Malta had been forced to retire to Alexandria because no air cover could be provided from Malta. The Spanish Civil War was very clearly imminent. The Italian naval effort might be confined to the Mediterranean but the German naval strength was concentrated around the three pocket battleships, the third of which was commissioned in January 1936. These ships were clearly designed for commerce raiding and could only be countered by the use of air reconnaissance.

There were, of course, aircraft in ships other than carriers. By 1936 there were sixteen aeroplanes in cruisers, and the aeroplane equipment of each capital ship was a movable feast, varying from time to time. The County class of cruisers were all reconstructed between 1935 and 1939, and the aeroplane complement uprated from one to three apiece, but the experience of war led to the aeroplanes and catapults being removed by 1943. The pilots of the cruiser aeroplanes are not listed as carried by those ships individually by name, but as on the strengths of Catapult Flights accredited to the local depot ships, for example HMS *Tamar*, the base depot ship at Hong Kong.

The aeroplanes carried by the carriers were as a rule inferior in performance to those of nominally the same function used by the land-based squadrons. This was not a question of the poor naval pilots being victimized by the wicked RAF. Aeroplanes for habitual use on carriers need, or needed in those days, to be of more robust construction than those operated on land. They have to stand up to the recurring stresses of being landed on heaving decks, and being manhandled across the same unsteady decks, in the open air or in the hangar. They are much harder to repair or to replace from the Aircraft Depot if they are damaged during a voyage. To allow them to be stowed they were often made to be folded up. All these factors meant extra weight, and added weight meant lower performance. Most Fleet aeroplanes during the period were specially designed or built for this purpose, and were not merely second-hand RAF types. However, when it was decided to adapt the Hawker Fury for carrier use as the Nimrod, the re-design brought the wingspan from thirty feet to thirty-three-and-a-half feet, and the weight from 3,490 lb to 4,258 lb: using the same engine, this cut twenty-five miles an hour off the speed and a thousand feet off the operational ceiling.'

Although the Admiralty announced the formation of the Fleet Air Arm in 1938, the change-over was not instantaneous. As late as September 1938, the carriers in commission, which did not yet include the new *Ark Royal*, all had their RAF headquarters personnel still in post. About a third of the pilots in the carrier squadrons of the Fleet Air Arm were RAF officers, and there was even one RAF officer among the catapult flight pilots. The last useful lists of carrier pilots come in the *Air Force List* of March 1939. *Eagle* was acting as a depot, and neither *Hermes* nor *Ark Royal* appear to be active. By now the carriers in commission have thirty-four RAF and seventy-nine naval pilots. *Glorious* now has a naval station headquarters: the other carriers still have HQs manned by the RAF.

Consequences
It is a truism that men do not react to the realities of a situation but to the situation as they perceive it. What the public and the politicians perceived was the impressive bulk of the carriers on the water. But the General Staffs had no illusions, or ought not to have had. Even the informed public may not have reacted in any positive way to the perceived image of the carriers. In 1925 the editor of Brassey's *Naval Annual* had commented on the rebuilding of *Furious* that it was strange to see so much money being spent on a non-combatant ship. The first genuine carriers for the US Navy, *Lexington* and *Saratoga*, used hulls built for two discontinued battle cruisers, and were left with the original secondary armament of eight 8-inch guns apiece. It was as if even US admirals could not believe that a carrier was a warship unless it had a respectable conventional armament. The British admirals could not swallow the fact that the 1,200 men of a carrier crew existed for no other purpose than to put a mere forty into the air, and if twenty of the forty were to be in a different uniform and dependent on an outside authority for their promotion and posting – it was more than they could bear in silence. *Ark Royal* had a larger complement than the contemporary *King George V*.

The great morale-building poster which was plastered on every hoarding in Britain in the summer of 1939 showed not carriers but the five *Royal Sovereign* class battleships in line ahead. The implicit message of the posters was that aeroplanes might be the main weapon of the Air Force, but for the Navy they were merely a useful accessory. Carriers did not gain face validity till after Taranto. But in 1939 they had ceased to be in any way an RAF responsibility.

The naval connection, at least the carrier connection, had

always been something minor in the eyes of the RAF. It soaked up only a very small proportion of the pilot strength, and for most of those it meant only an episode in their careers. Now that the connection was broken, and that slight manpower leak plugged, the RAF could devote itself entirely to being, in organization and activity, a land concern, an independent unit of the Army. But the Air Arm, with its massively expensive carriers and its vast demands for manpower, soaking up as many sailors as would a squadron of battleships, continued to be a vital concern of the Navy. Even if it was only a useful extra, it represented a large and necessary diversion of naval planning effort.

What the Admiralty thought their carriers should be used for can be seen by the general kinds of aeroplanes they armed them with. There were the types of carrier squadrons we have seen: spotter-reconnaissance, fighter and torpedo. There were no dive bombers, nor even any conventional bombers. The Navy saw the carriers as acting as units of a battle fleet, not as primary attack units. Their high speed would enable them to catch up after dropping out of the line to turn into wind and launch or recover their aeroplanes. Their reconnaissance aeroplanes would find the enemy, and then their torpedo aircraft would use their fish and, if not sink him, at least slow him down. Then the carriers would fall into their assigned place in the line out of harm's way, and their fighters would prevent the enemy carrier-borne or cruiser-borne aircraft doing the same thing in reverse. When the enemy was at last caught up with, the battleships would do the real business with their heavy guns. This may not be the main job the naval aircraft actually did in the war when it came, but it worked at least twice, in the chase of the *Bismark* and at Matapan.

Notes and References

1. A Boyle, *Trenchard*, Chapters 14-16. H.S. Roskill, *British Naval Policy between the Wars*. Both describe the stages of this dispute, but from different points of view.
2. C.G. Grey, *A History of the Air Ministry*, George Allen and Unwin, 1940, p. 77. This book has had a massive influence on historians of the Service, but close examination shows that it is based mainly on official handouts, and is valuable only for a few sidelights like this one. In

particular, Grey invented the term 'panic expansion' for the planned reorganization of the RAF after 1933.

3. A.J. Marder, *From the Dreadnought to Scapa Flow*, Vol. IV, OUP, 1969, passim, describes Jellicoe's repeated pleas for more air cover.

4. Marder, op cit, destroys the legend that the High Seas Fleet did not come out to offer battle after Jutland. See also AP 125, 1936 printing, p. 164.

5. *Conway's All the World's Fighting Ships*, Vols 2 and 3, Conway Maritime Press, 1980, neatly summarizes the Carrier data.

6. Brian Bond, *British Military Policy between the Wars*, OUP, 1980, pp. 24, 82, 97.

7. W.S. Roskill, op cit, Vol. I, p. 534, describes the effects of financial pressures on fleet development and on naval operating patterns.

8. H. Pownall, *Chief of Staff*, ed. Brian Bond, Vol. I, p. 83, Leo Cooper, 1972.

9. Owen Cathcart Jones, *Aviation Memoirs*, Hutchinson, no date (probably 1934). Jones was a Royal Marine officer, and clearly puts the social pressures on the air minded professional officer.

Chapter 8

Officers and Pilots

Definitions
The difficulties the Admiralty encountered in the 1930s in finding enough pilots for their aeroplanes were due largely to their insistence on having only officer pilots. What was the RAF situation, remembering that they offered to fill the gaps with airman pilots?

The key to the following discussion lies in Trenchard's concept of the General Duties Officer. As we have seen, he intended to recruit young men, via Cranwell or through short service commissions, to be officers, that is to perform the general tasks and duties which fall to the lot of an officer in the Army, or in any other armed service. In addition, these officers of the General Duties Branch, GD officers for short, would have to learn to fly, and this was a peculiarity of the RAF.

Today we can say first that many, but no more than half, RAF officers are pilots. We can also say that most, and perhaps all, pilots are officers. But in the 1930s we could have said that virtually all RAF officers were pilots, and most pilots, perhaps as many as three-quarters of them, were officers.

The *List* for May 1933 shows by name, 1,746 officers of the General Duties Branch (which is to say, qualified pilots) of the rank of squadron leader or below, as well as 145 students at the flying training schools not yet qualified. Of the qualified pilots, 1,001 are in squadrons or in flights with the Fleet: this includes the notional squadrons 15B, 22B and 24C, and the regular officers of the Cadre and Auxiliary Squadrons. Of the remainder, 498 are in posts at units where they may reasonably be supposed to be employed on flying, from flying training schools to the Anti-aircraft Co-operation Flight at Biggin Hill: there are at least nineteen of these small units.

There are also 114 GD officers at the various Group and Station Headquarters. They may be counted as among those who control the flying, and who need to be pilots to carry authority.

This leaves 133 GD officers in non-flying posts on stations. Twenty-one of them are on the course at Staff College, which has only two squadron leaders as the permanent staff, and fifty-three under training at the Home Aircraft Depot at Henlow, on the post-graduate engineering course which gives them the 'e' annotation. Sixteen are in the Apprentice Wing at RAF Halton, where perhaps they were required to fly the aeroplanes which the apprentices were supposed to build as a last masterpiece, and seven there in the Administrative Wing: thirteen of this total of twenty-three have the 'e' annotation. And the rest are in ones and twos in non-flying stations like the Packing Depot at Sealand.

There are other officers than GD officers. Leaving aside the Medical and Dental Branches, we have the Stores Branch with 209 officers, and the Accounts Branch with 112 officers in the relevant ranks (Table 8). Their distribution is, as one would imagine, rather different. Forty-nine stores officers and forty-three accountants are at the various headquarters, thirty-six stores officers and twenty-seven accountants are with squadrons, five stores officers are in the five aircraft carriers, and seventy-nine stores officers and eleven accountants are at the Depot at Uxbridge or Depot (Middle East) or with Stores Depots. There are several smaller units with these specialists, and the School of Stores Accounting at Cranwell has only two stores officers.

We may assume that the wing commanders and above in all Branches are in directing or commanding posts. But it would seem that the lower ranks of officer are not engaged in leading men: managing them perhaps in the case of the stores officers at the Stores Depots, but not in leading them. The traditional officer task was the leading of men, by platoons and companies and regiments, in battle, and the directing of the horny-handed seamen in working the ship. The GD officers are engaged in doing individual jobs of flying, using professional skills which they may or may not have been taught in the Service.

It is clearly no use trying to define an officer in historical terms such as his being a man who holds an office: if we want to know what Trenchard was trying to do when he set up an Officer Training Academy, or what the admirals and the generals meant when they said they wanted only officers as pilots in the aeroplanes allotted to co-operate with their units, we will have to look at a more social definition.

In the West today it is conventional to divide military men into two social groups. Each of these groups is characterized by its own traditional patterns of social intercourse, such as living, eating or, very important, drinking together, the use of certain speech patterns, by the sharing of off-duty recreations, and by marriage. These groups differ in the direction in which deference is paid and dominance exerted, and in which obedience is demanded. Passage from the lower deference group to the higher is possible and frequent according to well-known and formalized rules: passage from the upper group to the lower is rare and is seen as a matter of social obloquy. The groups differ in their general and military training, and in their daily working tasks. When men are needed for tasks requiring the exercise of high military or financial or political responsibilities, or of higher intellectual powers and attainments, they are drawn from the higher deference group. Members of this higher deference group are called officers.

Hopkins, in an otherwise sensible book, says, 'It was common knowledge that senior career officers frowned on the idea of commissioning aviators in the first place.'[1] It is difficult to understand where this idea comes from, unless from the German services where it was normal to use an NCO, or deck officer, pilot to fly an officer observer who was in command. Certainly in the British service what the generals and admirals wanted was that the aeroplanes for which they depended on their information should be flown by men who had already been given a measure of military or naval sophistication and should also be able to take the responsibility for the fate of armies, and therefore perhaps the nation. To get this level of sophistication and responsibility it was necessary to employ officers, although sometimes a very senior NCO, for example Mr Jillings, a Guards sergeant-major who was already qualified on paper to be a captain, could be depended on. Only as the reconnaissance task dwindled in numerical importance through the First World War, did it become acceptable to employ the occasional gifted non-officer recruit as a fighter or bomber pilot. It is said that Smith-Barry had to work hard to persuade the Americans to acknowledge this.

There is a continuing legend that the British officer corps of the First World War was composed of very young men just out of their public schools. Like most legends, there is a modicum of truth here. When the British Army suffered its earthquake in 1905, steps were taken to guarantee a supply of officers. What was going to be necessary in the second stage of any future war was a supply of young men able to do the business of a company. To ensure this they had to be of a secondary standard of education. In those days, secondary education, that is education above the level of the three Rs, was only available at fee-paying schools. The most reputable of these were the classical public schools which were boarding schools, but there were others. Most of the boarding schools had for many years organized units of volunteers or militia from their senior boys, since this was a convenient method of keeping a large number of boys amused and getting them tired out with the least possible effort by the teaching staff. Rugby was usually preferred to soccer in these schools for exactly this reason, it needed more boys in a team. In 1905 the War Office began to unite all these volunteer or militia units under the title of the Officers' Training Corps. A subsidy was paid, a syllabus worked out, and an examination set.[2]

The rule was that in time of war a successful graduate of this course, holding his Certificate 'A', would be entitled to a commission as of right if he had also a good report from his school and could find a colonel, any colonel, to countersign his application. The young gentlemen who flocked to claim commissions in 1914 may have been very young and wet behind the ears; but since they could put on their puttees, march and drill and fill in all the necessary forms, could site and lay out and pitch a camp, and could use their weapons, they were several steps ahead of their recruits. This explains how so many of the upper-class military figures of the period were able to take commissions as a matter of course. It also explains why the young Jack Slessor, seeking to join the RFC, had to pursue his

Major-General Sir David Henderson, as Director General of Military Aviation, wrote the Charter for the RFC in 1912, and in 1917 prompted Smuts to take it over for the first constitution of the RAF

(*Above*) Major-General Sir Frederick Sykes (*centre*), Henderson's deputy in 1912, was the second Chief of Air Staff and headed the RAF during Haig's hundred days of victories in 1918

(*Below*) The BE2c was the standard equipment in August 1914. Where would you use, or even stow, a rifle?

The Virginia, an enlarged version of the Vimy of 1917, and five knots faster, was the main bomber weapon of the service from 1924 to 1937

The Heyford, with a 30 knot speed advantage over the Virginia and a slightly lighter bomb load, began to replace it gradually after 1933, and would have faced the first Messerschmitts at Munich

The Anson came in 1936, and thus overlapped the Virginia. It was the first monoplane for front line service, but it was no faster than the Hind, had a smaller bomb load, and its range was only half that of the Virginia

The Wapiti, a later version of the DH 9a bomber of 1918, was being issued to Auxiliary Squadrons as a fighter as late as the mid-1930s

Instructors and pupils at 3 FTS going out to their Audaxes. During the 1930s, about half the Squadrons were equipped with Hawker Hart variants, yet they were still fulfilling the usual requirement for Wings that time be spent flying a 'service' aircraft

Aircraftman Shaw, the gentleman ranker, in the uniform designed for fighting the Boers

father till he caught him at the railway station entraining his battalion: it wasn't his father's signature as a father that the boy wanted, but any colonel's signature, and his father was the most easily available colonel.³

This was an effective way of officering the Army in a hurry. It was also an effective way of killing off an entire generation of men with secondary education. The Germans did the same, in one fell swoop, by mobilizing several divisions of senior secondary school boys under military age and throwing them into what was known as the Slaughter of the Innocents – *Kindermord* – at Ypres in 1915.

The system continued after the war. The RAF were able to base their entire system of officer training on it, and confine formal military training to airmen and apprentices. Short service officers did not have to be drilled. It was not till 1931 that Flight Lieutenant Hampton began to enquire what would happen in mass expansion when the public schools could not supply the entire demand. He outlined what became the sequence of wartime aircrew training, where the OTC syllabus was replaced by the Initial Training School.⁴

Sykes had suggested that officer training could be carried on for the independent Royal Air Force by the existing Army and Navy schools: that is to say, by Sandhurst and Woolwich, because Dartmouth was still organized for a twelve-year-old entry. For Trenchard this was not enough to develop and maintain a distinctive 'Air Force Spirit'. He decided early that his permanent officers must be trained at a new RAF College.

He chose a site for his new College at the old naval airship station at Cranwell in Lincolnshire. There were several factors in this choice. One was undoubtedly that Cranwell was a vacant station with a great deal of building already done, and all mains services, including the sewers, already to hand, even if the huts were largely temporary structures. Some of the naval brick buildings were still standing and operating well into the 1970s. One factor he would admit to was the distance of Cranwell from London, indeed from anywhere. The Cadets had to stay at the College almost all of the time. There was no taking the railway up to London to see a show and coming back in the small hours on the milk train, as happened at Sandhurst whether it were legal or not. The Flight Cadets must have envied Aircraftman Shaw who could make London for tea on a Wednesday afternoon by motor cycle, and be back for lights out. The College did hold on charge a stock of motor cycles which could be borrowed at weekends and for leave by senior Flight Cadets, and which were the frequent cause of comment by the financiers.

Trenchard kept the Directorate of Works under the Chief of Air Staff, that is within his own control. One of the major projects was the building of the College at Cranwell, a huge structure rivalling Greenwich, with a massive dome. The Queen's Colour for the RAF is housed there and has been there since it was bestowed, except for one or two unauthorized excursions to Gloucestershire. We may compare with the US Air Force Academy at Colorado Springs. Tourists at Colorado Springs or at Denver may take conducted bus tours to see around the Academy: nobody has organized conducted tours by bus to see Cranwell, but it might be a good idea if they did.

Trenchard organized a great deal of permanent brick building. Presumably this was as a kind of insurance policy, since if the Air Force owned a mass of well-built real estate, nobody would dare to abolish the service and have to find uses for the masonry, or even demolish it. This College Trenchard meant to be an academy for training officers. He saw it as a flying school only incidentally. Officers trained for an army have one basic skill in common. Whatever is their specialized role in armour or engineering or whatever arm they choose, they are all trained at first as infantry platoon commanders. Trenchard, whether he ever put this into coherent words or not, was in search of one basic role to be common to all his officers. Therefore he made it a condition of service that *all* of his officers should learn to fly. But flying he did not see as their primary job: they were above all to be officers, growing up to organize and operate a military service which happened, incidentally, to be armed with aeroplanes.

Trenchard, as we have seen, suffered from the delusion that flying is not a sport. To him, it was a serious business, as serious as war. He thought that young men joined a military service in order to become generals. But young men do not think that far ahead; they are taken up by what they see as the daily excitement of a military occupation. They did not before 1914 take commissions in the Royal Artillery because they were interested in doing mathematics or in firing great guns; they went to Woolwich because they wanted a life working daily with horses. In the same way young men did not go to Cranwell because they saw themselves one day as station commanders, if they, or anyone else in the 1920s, had any clear idea of what a station commander was or was going to be by the time they grew up to it. They went to Cranwell because they wanted to fly, every day, and for free.

But Cranwell was not free. Cadets at Cranwell, like cadets at Sandhurst or at Woolwich, paid fees, or at least their families did. There is a very real way in which one can say that these officers bought their commissions. They were not as well off as the regimental officers before 1870: then the young gentleman who laid out his money on first appointment as cornet or ensign did so knowing that if he survived he could always get his money back by selling out. But fees at Cranwell, as at Sandhurst, once paid were not returnable. You could only get a permanent commission before the Second World War by laying out a fair sum of money. The selection process in terms of medical examination and interviews may not have been rigorous by modern standards, but if you hadn't got the money to pay the fees you just could not go there. You needed to be able to put down a non-returnable cash payment of £150 on entry, which was in theory supposed to pay for your uniforms and equipment; and after that there were fees of £100 a year for two years. The down payment was certainly a large sum for those days.[5]

Of course, this element of cost dominated entry into any of the professions. In the period between the wars you had to be able to pay for your secondary education and then for your tertiary education, usually at a university. There was only one short period, of about a generation, between the 1950s and the 1980s, when it was thought just and practicable that ability should be the only qualification for a tertiary education and entry to a profession, and that all the costs ought to be paid by the State.

There were, of course, no recruiting advertisements for Cranwell, or come to that, for apprenticeships. *The Times*, *Flight*, and other sections of the respectable and interested press carried official announcements of each half-year's competition. It was always pointed out that a two-year course at Cranwell could be more economical – that is, cheaper – than three years at Oxford or Cambridge. That was quite true, even though the Oxbridge fees always quoted were those given in the relevant *Students' Yearbook* as what it was possible to get by on if you entered at a very undistinguished college, read an arts subject with no laboratory fees, and lived the quietest of lives possible, dispensing with luxuries like eating. It was probably not really possible to get by at Cranwell on the fees alone, but Cadets were attested as Airmen and paid as AC2s. The entrants to Cranwell could usually have afforded to enter any other professional occupation, such as accountancy or engineering or the law, but it was flying they chose, and that in spite of the fact that as an embryo lawyer or civil engineer they did not run about a one in five chance of getting killed before they had made any headway, let alone a 50 per cent failure rate in training, as we shall show.

Cranwell was always very small as compared with Sandhurst, although it probably cost the Exchequer as much, thanks to the flying commitment. Trenchard always meant it to be small and explained why in a famous speech. He had been, he said, a captain in the Army for twenty-one years. This was, incidentally, just not true: he had been a captain on his Regimental List for twenty-one years, but almost immediately he got his captaincy he was made by brevet a major in the Army, though not in his regiment, and during the rest of the twenty-one years he had gone up by acting and brevet ranks to be a major-general. But he had seen enough of men who had stuck at relatively junior officer ranks and mouldered out their lives. He was intent that such a thing should not happen in the Air Force he was building. He did the classical organization calculation. He had a plan for the expansion of the service over at least twenty years. He could therefore calculate the number of senior and air officers he would want in twenty years' time, and on this basis he planned the size of Cranwell. He was planning for a set number of officers who had been made suitable by background and training to hold posts of military or financial or political responsibility.

In 1922 the Chief Instructor at Cranwell, Squadron Leader C.F.A. Portal, described the training sequence. He managed to describe the Gosport system without ever mentioning Smith Barry by name. He enunciated two guiding rules. A cadet was started on aerobatics early, but there was no great stress laid on going solo. Solo times were in the range of from seven to twenty hours and were irrelevant to success. There was a better way of measuring the success of the training system. During the war a pupil was expected to break two aeroplanes up completely, and in addition to smash half a dozen undercarriages. In 1922 Cranwell was calculating on breaking one aeroplane and four undercarriages for every eleven pupils. The formal requirement was for seventy hours flying to wings, but in practice most pupils got in 130 hours during their two years.[6]

The half-yearly intake to Cranwell was planned at thirty cadets. Of the first two entries, thirty-one can be traced in the *List* as having been commissioned. That seems to mean that wastage in training at the College was approximately 50 per cent. This appears reasonable for a two-year course with a great deal of flying. After a further four years, three of this first year's output had been killed, one had resigned his commission, one was ADC to the Governor of South Australia, one was stationed at the RAF Motor Depot, two were with Fleet Flights, two with 4FTS at Abu Sueir (presumably as instructors), one with a cadre squadron at Aldergrove, and the remaining twenty-one were with operational squadrons. All these survivors were flying officers.

To these men Cranwell had been basically a flying school. After all, flying was the business of an air force. It was not till 1938 that the Commandant of the time, Air Vice-Marshal Baldwin, announced that he was altering the syllabus, to reduce the amount of time given to engineering and introduce the cadets to the business of the running of a squadron or a station.[7] That, of course, was what Trenchard had intended the whole place to do, with flying as a condition of eligibility to do this work: it had taken the RAF a long time to get round to it. Their restriction in experience was so great that for many cadets it was a surprise, on reaching their squadrons, to find that the majority of their comrades had not been through the College.

Temporaries

For Trenchard's calculation did not stop with the permanent officers. He was planning for a force of fifty squadrons within a few years, certainly before he retired. These represented a large number of aeroplane seats to be filled, far more than he could supply from the Cranwell output. Trenchard did not want his messes to be filled with ageing officers, who could no longer operate effectively but who were still on life-time engagements and would have to be paid and fill up the promotion lists. He took a drastic step.

Up to this time, the normal concept of a military officer was of someone who had entered the service, whichever service, for life, or until he chose to leave it: or until he was court-martialled and cashiered or dismissed, but in polite society one did not see that as a possible end of a career. In time of war, a gentleman might offer his services to the Crown for as long as hostilities lasted, but that was not a normal feature of life. The temporary officer serving on a contract with a fixed termination date was an oddity, unusual and of doubtful respectability. Trenchard made it usual and normal.

To fill his aeroplane seats Trenchard had three possible sources. First he could call for volunteers from among serving airmen. This source was limited: even if large numbers volunteered, he could hardly strip the servicing organization to provide pilots to fly aeroplanes which could not then be maintained. Secondly, he could, and did, call for volunteers from among the officers of the Army and the Navy. The response here was bound to be limited because of the reluctance of the other services to let their officers go to fly even for a short time, and by the limited use that could be made of these pilots: the Navy was willing to send some officers to be pilots in Fleet flights, and the Army would agree to let a few officers go to be pilots in Army co-operation squadrons. But that was all.

So to fill the mass of vacant aeroplane seats, Trenchard was forced to go public, and call for direct entrants from civil life. These men would be offered, in the delicate terms of the regular announcement in *The Times*, what was virtually a ten-year contract. It had been normal to recruit other ranks like this, but not officers. The officer entrants would be required to undertake to serve for four years active, and then to remain on the reserve for a further six years. Thus Trenchard hoped to solve the problem of the supply of replacements for casualties. Before we ask whether this was a realistic hope, we may note the terms of the announcement which stipulated that candidates must be British subjects by birth and of pure European descent. Such a formal requirement would raise a scandal in the 1990s, but at that time was taken for granted. At this date the coloured population of Britain was tiny, if not virtually non-existent, outside a few large ports. Further, the public order role of the military forces was still a shadow in the background of the services' thinking: it had always been a principle that the forces at home should never be seen as a force of black or white troops with black officers, available to put down riots or break strikes.

It was also assumed that all applicants would come from the social class which filled the public schools. In those days it was normal to speak of class differences openly. Captain Guy Pollard, VC, DCM, MC, writes as late as 1934, 'Fortunately in this country we have a large field from which to pick. The policy pursued by our Public Schools forms an ideal foundation for airmen in embryo. From early youth they are taught self-confidence and initiative, qualities essential to success . . . I do not wish to infer that Public Schools form the only recruiting ground, or that they necessarily supply the best pilots. There are many good sergeant pilots promoted because of their ability. But they are naturally endowed with those attributes which others more fortunately situated in life acquire by education.'[8]

We must note that this language was normal for the time. There was no explicit statement here restricting recruiting to one educational or social class, simply the bald assumption that only members of that class will apply. There is also the interesting assumption that the powers of leadership seen in that social class are merely the result of education and not inborn in them.

How far was this scheme successful in raising a pilot force or in creating a reserve? It is a simple matter to trace the early entry to the Short Service Scheme by name through their four years service. In the *Lists* of 1923 the names were traced of officers entering the Flying Training Schools. They would have spent first a short period, usually two weeks, at the RAF Depot at Uxbridge, getting their uniforms made and learning whom to salute: any other military training they would have received already at their public schools. From the Depot they would then be posted to a flying training school for approximately a year's pilot training. Allowing for leave they would be ready for posting to a squadron after about fifteen months' service. If they survived.

The names examined for 1923 were the first 125 alternate

names on the list of officer entrants to flying training schools. Of these, twelve were killed within their first four years service, one died of injuries and one is simply recorded as dead. Seven of these were killed within what would seem to be the flying training period. Squadron Leader Portal did not give a death rate for Cranwell training. Thirteen relinquished their commissions on grounds of health, but only three within the training period. Six had their commissions terminated within the training period, and one must guess that they had failed to learn to fly. Sixteen resigned their commissions, of whom eleven left within the training period. Two were dismissed after being court-martialled, and one is simply recorded as being removed from the list as 'His Majesty has no further use for his services.' This means a wastage of nearly 25 per cent within the flying training period.

Altogether sixty-three, or 50 per cent, are still on the *List* in May 1927. Three of these have already passed to the Reserve of Air Force Officers (RAFO). Of the remaining sixty, sixteen are in posts which do not strike one as flying posts, such as one man at an Armoured Car Squadron in Iraq, and one on the staff at the Depot at Uxbridge: there are three on the administrative staff of the Apprentice Wing at RAF Halton. Six are under training at the Depot at Uxbridge, and one, with no explanations given, at Depot Middle East. And two are no longer on the General Duties Branch List, but are shown on the Stores Branch List. All these men in ground posts had originally been shown as posted to squadrons. One can only assume that they had not done well in their first period at the squadron and had been removed from flying duties, to serve out their time elsewhere. In modern parlance they had failed to fly at the Operational Conversion stage. Sykes, it will be remembered, had assumed that all staff and administrative posts could be filled from such men who were no longer, for one reason or another, fit for flying employment.

Of those who were still in flying posts, thirty-two are in squadrons and another three at the bases at Gosport and at Calshot. Three are with the Experimental Unit at the Royal Aircraft Establishments. Nine are flying instructors, two at Cranwell, five at FTSs in the UK, and two at 4FTS in Egypt. And one, significantly, is under training at the Home Aircraft Depot at Henlow. He was on the course which led to the award of the 'e' symbol: entrance to the course was by a competitive examination, and normally carried with it a permanent commission.

Commissions, Posts and Messes

The strength of a flying unit would therefore include three classes of pilots, on three different kinds of engagement. There were sergeant pilots, usually, but not always, after 1923 ex-apprentices, who were committed to flying for five years and would then return to their basic ground trades. They lived in the sergeants' mess. There were short service officers, who had originally engaged to fly for four years, and would then return to civil life with 'reserve' against their names for another six years, unless they were awarded permanent commissions, or were allowed to extend their engagements: otherwise they had little prospect of promotion beyond flight lieutenant. And there were the Cranwell graduates who had on completion of training been granted permanent commissions, and would serve till the end of their active working life; they had guaranteed prospects of promotion and might reach Air Rank. All these men were doing the same job, as squadron pilots.

Also in the officers' mess were officers of the other Branches, the medical officer and the chaplain, and also the officers of the Stores Branch and the Accounts Branch. They were on varying engagements, and had some promotion prospects, varying with Branch.

The greater number of the inhabitants of the sergeants' mess were old hairies, who had risen over a long period of service to their three stripes, or even to Warrant rank. They differed greatly in their interests and way of life from the young sergeant pilots.

Let us return to the earlier definition of military society and of the officer, and ask, why is this military society divided into only two groups?

When Trenchard began to remake the Air Force, there was no obligation on him to retain the two-mess structure of the Army. Army messes, of officers and of NCOs, had originated when the British Army had first gone to war in an undeveloped country, Spain in 1700, and officers found it more convenient to club together to hire servants and a cook and bulk-buy food, instead of each living on his own, turning up for duty but finding quarters and meals in an inn, as they were used to doing in Britain and in Flanders. The mess system played a part in developing the esprit de corps of the later County Regiments, but its principal merit was in reducing the cost and trouble of catering. Size in a mess was on the whole an advantage.

But the Navy had a very different system in origin. A ship is made of a number of small boxes, and a mess is most convenient if its membership is no greater than will fit conveniently into such a space. In the old Navy, certainly down to the 1920s, a notional ship had at least two officers' messes. One was the Wardroom, which was the living space originally of the 'Commission Officers' of what we now call the Seaman Branch. They shared this with the two members of the complement who were socially of gentleman status, the doctor, who was well paid, and the chaplain, who was paid very badly and in vast arrears but who, as a university graduate, was a gentleman and could not be left to sling his hammock with the common seamen. The Captain was not a member of the Wardroom but messed by himself. The occupants of the Wardroom therefore included for the most part men who had arrived in it after some service and who looked forward to moving up out of it.

There was a more junior officers' mess, called the Gunroom. In this there lived the three Warrant Officers, who were permanent parts of the ship's establishment and stayed with her when she was out of commission or, as we would say now, in mothballs: they were the Gunner (armament technician and keeper of warlike stores), the Carpenter (shipwright) and the Boatswain (motive power department). They had usually, but not always, worked their way up from the ranks. There were also other specialists, in particular the Master and his Mates who were navigators and usually entered direct from the merchant service, already trained and mature. The history of the Navy over three hundred years is largely that of the struggles of each of these classes of Officers, or Branches, to enter the Wardroom. And there were the very junior members of the officer class, the Midshipmen and what were successively called Passed Midshipmen, then Mates (not the Master's Mates), then Sub-lieutenants, who were qualified on paper and by length of service for Lieutenants' posts and who were simply waiting their turn in the promotion ladder. So the Gunroom accommodated men who had attained it after long service and did not expect to rise any further, men who had entered it direct with no service on the lower deck and did not expect barring an administrative revolution to go any higher, and men who had also entered it direct but saw it as a stage before they passed up beyond it.[9]

The old German Navy had an even more complex system. There were officers of the Imperial Navy who were few in number, were if not aristocratic at least socially on a par with the Army officers, were the Command Branch and did not deal with the men; and the 'Deck Officers' who shared their mess on rather inferior terms, who actually ran the ships, were usually of middle-class origins, had direct contact with the seamen, and who had no prospect whatever of command.[10]

These alternatives must have been known to Trenchard, especially since some of the most senior officers of the new Royal Air Force had originally been naval officers. But he chose the two-tier system he had been used to in the Army. This system was to create vast problems later in the century, when there were large numbers of pilots, and the policy on commissioning, if there were a policy, was not immediately obvious. It is tempting to imagine that these problems, or some of them, might have been evaded if a Gunroom had been adopted, in which one would have found junior officers of all Branches and also the warrant officers and sergeant pilots.

As the Second World War went on, many stations adopted the system of a single aircrew mess regardless of commissioning status.[11] After that war, the problems were eventually solved by the expedient of commissioning all pilots regardless of their entry method, under the slogan of 'if a man's good enough to be a pilot, he's good enough to be an officer'. Trenchard in 1920 was saying exactly the reverse, that if a man were good enough to be an officer, he must also become a pilot. When the original decision had to be taken, in 1920, it seemed more important to have a status system which was compatible with that of the Army, since the Air Force was seen as likely to work in closer collaboration with the land forces than with the Navy.

Branch
In the old Army mess of Trenchard's youth, virtually everyone was there because he was a combatant, whose job was to

command a sub-unit in action. The exceptions in an infantry regimental mess were the Quartermaster and the Paymaster. The former was normally a ranker, near the end of his service, and given an honorary or titular commission to go with the job of storekeeper and to raise him, for his honour, above petty peculation. The Paymaster was usually a retired or invalided officer, re-employed. A regiment, or later a battalion, in a foreign station would often have a Royal Engineer officer attached. In other words, the Army was a world of generalists, with only an occasional specialist. In contrast the navy was a world of specialists, of whom only one group of experts were there to steer the ship in action and take military decisions. If the Navy wanted to have good men to spend their whole careers in these specialist branches, engineers and pursers in particular, they would have to offer them promotion prospects divorced from ship command and protected from outside competition.

An officer in the Navy, therefore is commissioned in a particular branch. In this branch he will remain for his service, knowing the limits of his responsibility and also knowing that his promotion prospects will not be threatened by competition from generalists – that is from the executive or seaman branch. The *Navy List*, therefore, is a number of seniority *Lists*, one for each of the Branches. The promotion ceiling in each specialist branch is limited but high, usually to Flag Officer rank.

In the Army, in theory, an officer is commissioned into a particular regiment or corps and remains in that regiment or corps till he retires or till three years after he is promoted to full colonel. He then becomes a colonel, late infantry or whatever, and hopes now to progress to General Officer rank. In the British service there is no such animal as a General of Artillery; there are only generals, each, regardless of his earliest training, capable of appointment to command of forces of all arms. Experience during the First World War of the employment of brigadier-generals, commanding infantry or cavalry brigades, led to the renaming of this rank as Brigadier.

In the American service there is simply a seniority list and officers are qualified for particular posts by virtue of their military specialty. At the outbreak of the First World War, the Commanding General of the US Army and his deputy both happened to be medical officers by military specialty.

The point at which Trenchard made a conscious decision as to which method he would follow may be seen from the *Lists*. In the first *Air Force List* most officers were on a common seniority *List*, except for the officers of the Stores Branch, who were counted separately, and for the officers of the Medical Branch below the rank of lieutenant-colonel. The more senior Medical Officers were on the common *List* with all the General Duties Officers, that is the officers who were not restricted by their training to any particular specialty but who were qualified for command, or for a post of military or financial responsibility. Officers on this list were annotated for particular specialities, such as balloon or airship pilot or medical officer: but all were seen as competing for promotion to more senior General Duties posts. This looks very like the American system.

By the following year, the new rank titles are in use; the *List* had shrunk and also stabilized. The Medical Branch was separated from the General Duties Branch right up to the top; in other words the doctors were having no competition from outside for their promotions. There was a tiny Stores Branch, in which the senior officer was a wing commander, none other than H.H. Kirby, VC, and a Stores Accountant Branch, even smaller (Table 8). These officers were also to be given promotion channels protected from competition from the generalists – that is, if a post were annotated for a Stores Branch officer, there was to be no posting of a GD officer into it, but a Stores Branch officer would have to be posted or promoted into it when it was vacant. There was a Legal Branch, which was usually countable on the fingers of one hand, and a Chaplains Branch. And that was all.

There was no Engineering or Technical Branch. Engineering posts were filled by pilots who had been through the course at the Home Aircraft Depot at Henlow, to earn the 'e' annotation. Posts for photographic or armament or signalling specialists were filled in the same way by pilots who had been through in-service training courses and earned the relevant symbol. Navigation specialists were pilots who had been through the Air Pilotage School and earned the 'n' annotation.[13]

There was no Administrative Branch. Administration at station or higher level was carried out by pilots on ground tours, or at the lower level by squadron pilots as a secondary duty.

There was no Education Branch, but pilots were at least not expected to teach airmen in their spare duty hours. There was an education service, of civilian graduate teachers serving on good civil service terms, of officer status – and therefore, like the old naval chaplain, members of the officers' mess – and holding reserve (Class C or CC) commissions. This may have been a better arrangement than the Army made when an Army Education Corps had been formed from scratch in 1919. Applicants were interviewed and if satisfactory were given commissions in ranks dependent on their war records as fighting soldiers and taking no account of age or academic record or teaching experience or previous seniority: as a result there was a total promotion block, and when the Second World War broke out most of the officers were serving in the same ranks as they had originally been given in 1919.[12] There was no Air Traffic Control Branch because there was no air traffic, and the circuit was quite effectively controlled by a Duty Pilot in the tower, with a Very pistol, and with a pair of fieldglasses if he wished for them. And no Fighter Control Branch because warning and interception and air fighting were done entirely by the naked eye.

We must mention here that the RAF was given a Staff College in Trenchard's great scheme. The Army Staff College shared a site, and almost a building, with the main officer training establishment: the RAF Staff College was put down in a convenient building, at Andover, where it was a lodger unit on the station. The main task of the Army Staff College was to train staff officers for minor staff posts, and the main diet of training was schooling in making out march orders and routes. In the same way the RAF Staff College trained officers mainly in the making out of Task Charts. In making out a task chart, for a flight, a squadron, a station or whatever, you start with the job the unit is being 'tasked' to do: this may be something routine like carrying out flying training, or it may be something special like a single bomber raid on an important target. The beginning is to work out how many hours flying that will account for in preparation and in execution, and over what period. The end result is a detailed account of what this will entail in terms of aeroplanes, spares, fuel, ammunition, accommodation and men of every possible trade, from pilots through fitters to cooks and waiters and aircraft hands general duties: and, of course, for passing-out parades or a victory celebration, the Band. Making out a task chart is the very stuff of air planning and requires more subtlety than at first appears. Nothing must be left out, or left to chance. And the great war plan of 1934 on which all the expansion was based was, crudely, little more than a vast task chart for the whole Air Force, based on a particular job, fighting the Germans, at a particular time, spring 1939.

And that, in 1933, when Trenchard had retired and the Salmond brothers had succeeded each other as Chief of Air Staff, was that. It was a Pilots' Air Force, organized in squadrons which in turn were set up like miniature battalions with officers and men as integral members, with Colours and badges and mottoes and sergeant-majors, and in all its internal organization and social structure very obviously an offshoot of the Army. The service had had thirteen years to make its system of command and control and supply into something sensible which men could operate with their eyes shut, or in the fog of war which is much the same thing. The aeroplanes were not much beyond what had been around thirteen years earlier. And as for a strategic philosophy and plan – there was none because there was no obvious or real enemy for an Air Staff to plan against. Then, in the summer of 1933, everything hit the fan. And hardly anybody outside the Air Force noticed.

Notes and References

1. G.E. Hopkins, *The Aeroplane Pilots*, Harvard University Press, 1971, p. 14.
2. A.R. Haig Brown, *The OTC in the Great War*, Country Life Publications, 1915, passim, but esp. p. 80. This may well be the first use of the title 'Great' War for the conflict of 1914–18, the term having been used up to then by military historians for the Peninsular War.
3. J. Slessor, *The Central Blue*, Cassell, 1956.
4. Flt/Lt. H.N. Hampton, 'The Training of Pilots for the RAF on the outbreak of War', *RAF Quarterly*, April 1931, p. 226 seq.
5. R.P.M. Gibbs, *Not Peace but a Sword*, Cassell, 1943, p. 15 on the burden of Cranwell fees or the middle-class family of the time, and p. 36 on the shock to many Cadets meeting non-Cranwell officers in the squadrons.
6. Sqdn/Ldr. C.F.A. Portal, 'Methods of Aeroplane Flying Instruction', *Aeronautical Journal*, May 1922.

7. Flight 23/12/37, announcing AVM Baldwin's policy for the coming year, notes that this, with an increase in the number of sergeant pilots, marks a move away from the concept of the GD officer as a pilot to the concept of command. Which was where Trenchard came in.

8. Captain A.O. Pollard, VC, DCM, MC, *The Royal Air Force: a Concise History*, Hutchinson, 1934, p. 242. This gallant author of a history of the RAF remarks in a puzzled way that many of the RFC pilots in 1914 had had some Army training.

9. For this development, see Michael Lewis, *England's Sea Officers*, Hodder, 1939, passim: Charles Morgan's novel, *The Gunroom*, 1919 (reprinted Chatto and Windus, 1939), esp. p. 18 is informative on the social position of the Engineer Officer in the Royal Navy before the First World War.

10. H.H. Herwig, *The Imperial German Naval Officer Corps*, OUP, 1970, Chapter 6. The French Royal Navy before 1789 had a similar system.

11. Lord Tedder, *With Prejudice*, Cassell, 1966, p. 185.

12. A.C.T. White, VC, *The Story of Army Education*, Harrap, 1963.

13. J.A. Kent, *One of the Few*, William Kimber, 1972. This is a fascinating account of the origin, attitudes, training, work and (especially on p. 59), career prospects of the short service officer. Allen Wheeler, *Flying between the Wars*, G.T. Fowles, 1972, has a good account of pilot training but opens with a revealing statement by the then adjutant at Digby on the in-service reputation of short service officers.

Chapter 9

The Great Change

A New Situation

What broke up the world was Hitler. The textbooks usually say that Hitler came to power in 1933. It was not quite as simple or clear cut as that.

For some years before 1933 the Nazis (short for the National Socialist German Workers Party) had been one of the major parties in Germany, a state where there were a multitude of political parties. In the presidential election of 1932, Hitler, with eleven million votes came second to Hindenburg, the hero of the First World War, with eighteen million. The Communist candidate got five million. In the following general election of July 1932, the Nazis were the largest single party, but did not have a majority and would not enter a coalition. A second election in November of the same year left the situation unchanged, but Hitler was now able to become Chancellor in a coalition with other right-wing parties. A third general election was set for March 1933, and during the campaign the mysterious burning of the Reichstag building allowed opportunity for a right-wing dirty tricks campaign; most features of what was then considered disgraceful behaviour in an election are today regarded as only mildly discreditable. The Nazis, with their Nationalist Party allies, gained a bare majority in the March election, and formed a government. In late March, this assembly passed an Enabling Act giving the government dictatorial powers for four years. In May all other parties than the Nazis were, in one way and another, abolished, and in a general election late in the year the Nazis gained 92 per cent of the popular vote.

The whole process of gaining power thus covered about the same space of time as an American presidential election campaign, and Hitler's period of unchallenged power with total military and political success covered about the period of two American presidential terms; of Roosevelt's first two presidential terms, in fact. The effect of the Battle of Britain was to end this run. Although there was never an overnight *coup d'état*, things happened swiftly enough in detail to give that impression, and induce some slight surprise in the survivors as to how short a time the whole episode had taken, considering how much damage was done to the world.

This must be seen against a background of attempts by politicians, well intentioned if ineffectual, to prevent the damage. The League of Nations had in 1930 called for a general Disarmament Conference. This opened in 1932: sixty governments attended, including the United States (not a member of the League) and Germany. The second session, in 1933, ended when Germany with some effect withdrew both from the Conference and the League.

In March 1935, the Germans denounced the clauses of the Versailles Treaty which imposed German disarmament, and introduced conscription, as well as formally founding an air force. For most of the inter-war period, a German air force had been developed in the form of a cadre secretly trained in Russia. The Germans occupied the Saar in March 1935, and took over Austria in March 1938. It appears now that the German Army was not, in the invasion of Austria, the highly efficient invasion machine it had been in 1870 and 1914 but it was, of course, just working up.[1] It seems likely that the most effective German military force at this time was Brigade Thaelmann, in Spain. At the Anschluss the commanders of the Communist Brigade Thaelmann petitioned the Spanish Republican Government that the anarchist Austrian Battalion Dolfuss should be brought under their command, since 'we are all Germans now': the ensuing course of relationships between the German and the Austrian comrades did not run smoothly.[2] Foreign politics is seldom simple.

The first Czech border crisis took place in March and April

1938, and the second Czech crisis, usually known as the Munich Crisis, in September 1938. The third Czech crisis was in March 1939, when the Germans occupied most of Czechoslovakia and were presented with the *matériel* of the Czech Army. The full military alliance between Germany and Italy came in May 1939, and a non-aggression pact between Germany and Russia was signed in August 1939.

In the Mediterranean, the Italians had invaded Abyssinia in October 1935, after a run-up of several months in which the League of Nations tried to get support for economic sanctions. At the opening of this war, the British Fleet as already described assembled at Malta in some strength, but was forced to withdraw to Alexandria because of the Italian air threat. The Spanish political scene was disturbed from early in 1936; formal hostilities opened in June and continued till March 1939. As late as 1973, Shinwell was of the opinion that we should have mobilized and threatened war in 1936.[3]

The United Kingdom
In Britain in 1933 the memory of the First World War was still too fresh to make thought or talk of war welcome. The nations of Europe had gone off to the First World War as if eager for it, but the Western democracies were not willing to undergo the same experience again unless compelled. The British Prime Minister from 1931 to 1935 was Ramsay MacDonald, leading a coalition of Conservatives with dissenters from the main Labour and Liberal parties. MacDonald himself was a pacifist: in 1935, he remarked to the Chief of Staff, warning him of coming war, that 'this won't be the first time the services will have entered into a war unprepared.'[4] He allowed his deputy, the Conservative leader Baldwin, to act for him as head of the Committee of Imperial Defence. After the Conservative victory in the general election of June 1935, Baldwin became Prime Minister; he was therefore concerned in, if not responsible for, every major defence decision of the decade. Already in March 1935 the Government had announced that Britain would have to put her house in order as far as defence was concerned. After the election of 1935 a programme of re-armament was initiated although it had not been a plank in the Conservative election platform.

In May 1938 Baldwin was succeeded by Neville Chamberlain.

He was the main mover on the British side in the Munich negotiations of late 1938 and in the decision to go to war in 1939.[5]

In March 1933 the Chief of the Air Staff was Marshal of the RAF Sir John Salmond; he was succeeded in 1934 by Air Chief Marshal Sir Edward Ellington, and he again in 1938 by Air Chief Marshal Sir Cyril Newall. Air Vice-Marshal H.C.T. Dowding became Air Member for Supply and Research in 1931, and held the same post, with a change of title to Air Member for Research and Development, till he became Commander-in-Chief Fighter Command in 1937 as an Air Chief Marshal. Other names will be mentioned as they come up, but these are important.

State at the Start Line

We may therefore consider the summer of 1933 as the starting date for active preparation for a war against a known enemy. At each later stage of crisis, the politicians wrung their hands and dithered; but from this date, the Air Staff simply shrugged their shoulders and went about doing what had to be done. A first plan for expansion was presented to the Committee for Imperial Defence in November 1933, and work on it cannot have started later than the summer of that year.

It can be assumed, then, that at that date the Air Staff were asked the crucial question: if war breaks out tomorrow, how many sorties can we launch immediately, and with how many aircraft? The main concern of course was defence of, and attack from, the home island.

The establishment of aircraft at this date was twelve per squadron of fighters and light bombers, eight or ten per squadron of heavy or night bombers, and four or five per squadron of flying boats. The *List* shows that the normal fighter or light bomber squadron had on strength something between sixteen or twenty pilots, of whom five or six would be sergeants. The variations between squadrons are quite wide. The heavy or night bomber squadrons would have usually twenty-four pilots, with about nine or ten of them sergeants: they would usually fly with two pilots per machine and one to three air gunners, normal servicing tradesmen who did this job as a diversion between spells of their usual work and were paid an extra sixpence a day to be available. The Army co-operation squadrons in the UK would have about twenty-two officer pilots and no sergeants. The flying boat squadrons in the UK varied in strength between ten and fifteen pilots, about one third of whom were sergeants.

Not all the pilots shown would be effective. Accounts from the period show a fairly uniform practice. A new pilot officer or sergeant pilot posted in straight from the flying training school was not usually considered as a permanent asset or as fit to fly in action. He would be given a more or less planned period of training on the squadron, lasting for three or four months, sometimes longer, according to progress. The degree of planning was not always apparent to the student. During this time he was on probation, and might at any time be judged as incapable of employment and returned to the Depot. If he were thought satisfactory, he would at length be taken aside by the squadron commander and told to go and buy himself a squadron tie, and sew the squadron badge on his overalls. This period of post-graduate training usually ended a little while before the pilot officer had completed the two years service necessary before promotion to flying officer. We may therefore consider the effective pilot strength of the squadron as consisting of the flying officers and above, and a proportion, say about half, of the pilot officers. The sergeants would give seven years active squadron service after training, and we may therefore make a guess that each squadron might have on average half a sergeant still on probation.

In the United Kingdom in May 1933 were three squadrons of Hawker Furies, nine of Bristol Bulldogs, and one of Hawker Demons, the latter a two-seater machine originally known as the Hart Fighter. All, although they had about fifty knots extra speed, would have fitted very well into the 1918 line up. This meant a theoretical aircraft strength of 156. The fighter pilot strength was, by a count of names, 208, which allowing for eight pilot officers out of sixteen and seven sergeants out of sixty-seven as not yet trained would give 193 effective. Perhaps one could imagine a pilot in conditions of desperate need managing three sorties in one day, but not on two days running.

There were seven squadrons of light or day bombers available, four of Harts and three of Gordons. These aeroplanes were recognizable to the naked eye as descendants in structure and role of the DH9a, which itself was only recently out of service. The Wapiti, built largely from unused DH9a spares by Westland (who had been the principal contractors for building the DH9a), was still in use. These squadrons had 137 pilots, of whom seven pilot officers and four sergeants may be counted as not yet effective. Again it may be possible to think of these units as capable in emergency of two sorties per effective pilot on the first day. But not two days running.

We may therefore think of the Air Staff as being able to tell the politicians (or the German Air Attaché being able to tell Göring) that on the first day of hostilities they might, with luck and low casualties, expect nearly 600 fighter sorties and 250 bomber sorties, but never more than 150 fighters or eighty bombers in the air at the same time. And that only if the bomber targets were already on the island. If the bombers had to make a flight to the coast and then refuel, it was unlikely that more than one sortie would be possible in the day, and then only if the enemy had already reached the mouths of the Scheldt which was about the limit of range of the aeroplanes available.

For night operations, there were immediately available four regular squadrons of Virginias, one of Hinaidis and one of Sidestrands, all very recognizably near relations of the Vimy of 1918. They might get in one sortie each during the hours of darkness, say fifty aeroplanes in all, which should allow them from their main bases on Salisbury Plain to attack the mouths of the Scheldt after a refuelling stop. However their speeds were very low, especially the Virginia, and their defensive armament small, so that it was doubtful if there would be any left to launch a strike on the second night. In any case, nobody had had any experience of flying over a blacked out countryside so that it was merely a matter of hope that they could find their targets. Germany was safe. If, as is sometimes alleged, Trenchard was personally responsible for the idea of mass bombing of civilians, one can only ask from the hard evidence of the *Lists*, what he thought could be done against any possible enemy with this location of bombers, of these types.

The politicians would then ask how it was that while they were paying the wages of 2,000 pilots, only about 400 could be available in time of war. What about the reserves which was what the short service scheme was intended to produce in large numbers? Things, the Air Council would have to reply, were not as simple as that.

On the *List* we find in May 1933 an item called the Reserve of Air Force Officers (RAFO): this was divided into three classes, of which class A, consisting of flying personnel who were RAF trained, is relevant. Class B were technical or professional specialists and class C were civilians employed by the service, who all had had RAF training: Classes AA, BB and CC were parallel but were men with no service experience. Class B and BB were very small, while the members of class C were mostly career civil servants employed abroad, like the Civilian Education Officers, who might for convenience be mobilized in emergency for their own protection. Maurice Baring's honorary reserve commission was in class B.

Each man is shown with his date of entry into the reserve. Therefore those officers with reserve entry dates of less than six years earlier would include all short service pilots who had gone to the reserve. In May 1933, this group stood at eighty names. All these men would be in theory liable to recall and would have been under an obligation to return each year for refresher flying training. It is possible that some of them might now be in occupations, under the Crown or otherwise, which would be officially counted as more important than reporting to the Colours. The eighty would therefore be a maximum. Even if these men had kept up their reserve obligation, at any time in the year half of them would be now out of flying practice and needing a fortnight or more refresher flying, and few of them could be guaranteed to be current on the types now in use. Although many of the types still flying in 1933 had been in use in 1927, in later years this consideration would become important. Certainly this reserve would not consist of the vast army of men able at once to join operational squadrons which the innocent politicians envisaged.

Immediate reinforcements for the squadrons to replace first day battle casualties would have to come from within the service. The two notional squadrons at Martlesham could contribute at once twenty-four pilots in flying practice on operational types. There were eighteen pilots under training at the Air Armament School and the School of Photography, also therefore in flying practice on operational types. There were also eighteen officers of squadron leader rank and below at the Staff College, and at the Home Aircraft Depot at Henlow there were fifty-three officers under training for the in-service engineering qualification. All these were in flying practice on modern types, having been posted from squadrons within the last year, and available immediately on the outbreak of hostilities. They would add nearly a quarter to the usable pilot strength in the first week of war. By the end of that week the cadre squadrons would be mobilized and available, giving two Virginia squadrons and one of Hyderabads, or thirty aeroplanes and fifty-four pilots for a night strike. They would also provide one squadron of Wallaces (an updated Wapiti) and one of Horselys, with thirty-five pilots, to provide an extra seventy sorties on probably the second day. The six junior Auxiliary squadrons with their obsolete equipment of Wapitis[5] could only be regarded as providing reserves of eighty-two pilots for other units; although two Auxiliary squadrons at Hendon, with Harts and with five regular and thirty-five auxiliary officers altogether, might be regarded as giving another eighty sorties on, probably, the second day.

Reserves for later in the opening stages were more problematic. The seed-corn principle applies: you cannot both sow it and eat it. By stripping the instructors from the home-based flying training schools and postgraduate schools, 183 pilots could be supplied at once.[6] But this would mean shutting down the flying training system. To look forward to a moderate increase at the end of a month and to a continuing flow thereafter, it might be more realistic to accelerate the training of the senior courses at the FTSs and at Cranwell. The First World War had taught the consequences of cutting pilot training too short.

In any case, the pilot effort needed would be too small. But to increase the pilot effort meant other increases. In a speech at the end of 1934, Baldwin said it was all right to call for more aeroplanes, but more aeroplanes were of little use unless there were aerodromes to land them on. It took about four and a quarter years to set up an aerodrome, the first half year being taken up with acquiring the land; there was no compulsory purchase. Even where airfields of the previous war were still in public ownership, the land was leased to farmers who would have to be given some notice. If there were to be any increase in the size of the Air effort, it was going to take some time. And it was worth considering what kind of Air Force ought to be built up.

The Plan

Baldwin was very conscious, as was every politician except perhaps MacDonald, that it was the small German expenditure on aerial bombing which had come near to bringing down Lloyd George's government in 1917: that had produced a separate Air Force. It might come near to bringing down a government again. In November 1932, Baldwin had made a speech in which he said, 'the Bomber will always get through'. This was not merely a prophecy of doom; it was a statement of an experimental fact. It was demonstrated in the Air Exercises of 1933.[7]

The RAF in those years had a formal training year, starting in October with individual training; the spring brought squadron exercises, and the summer and early autumn were given over to larger formation exercises, group, command and even whole-service manoeuvres.[8] The exercises of summer 1933 were especially elaborate and on a large scale, combining both fighter and bomber units. They were reported in full in the press. A large number of reporters were taken on the exercise in bomber aeroplanes, and saw it all.

The bombers flew by day and by night. The tactical doctrine was that the bombers should fly in close formation, to provide Boxes of fire from their flexible Lewis guns so that fighters could not penetrate from any direction without being engaged. These tactics had been successful in the days of the Independent Air Force in 1918. In the Air Exercises of 1933 they were judged to be successful again. The bombers claimed that they had shot down all, or almost all, of the fighters sent against them. The fighter pilots claimed to have shot down all the bombers. As each bomber carried from two to five men, against one man in most of the fighters, and the umpires and journalists were carried in the bombers, there were no doubts at the official level that the fighters were outvoted and the bomber tactic was vindicated.

A second important lesson was learned. The bombers *did* always get through. When they flew at a great height, even in daytime they could not be seen from the ground. They could be heard, but although the Army promised great results from their acoustic batteries, the fighters could not rely on the briefings they were given before they took off, or the corrections they were sent in Morse. In fact, the only definite evidence of the bombers' whereabouts came when they signalled to the ground, as they did in exercises, that they had hit their targets and what those targets were. In wartime, the only evidence would come when the bombs hit the ground; it was assumed at this time that all bombs dropped would hit their targets. The bombers could not be repelled, they would always get through and would always do damage.

The interceptions came on the way home. If the bomber could be located at the moment it turned round for the trip home to base, and if the location of the base were roughly known, the track could be projected and the fighters positioned for visual interception. At night, the fighters chased concentrations of searchlight beams, with greater success.

The lessons of these exercises, and of the parallel exercises of the following year, was that prevention of air raids by fighter opposition was virtually impossible. The only method, as *Flight* pointed out, was by attacking the enemy's bomber bases: or by what the old gunners Salmond and Ellington would have called counter-battery work.

Almost immediately afterwards, a second series of exercises, with similar regard for publicity, was mounted by the RAF with naval co-operation. The press briefings were at pains to stress that these were not Navy-Air Force exercises but Air Force exercises in which the Navy was co-operating. Reading through the press reports with hindsight it is clear that while the RAF carried out a series of bombing sorties, the Navy provided targets at sea off the north-east coast of Scotland. The exercises tested two sets of theories. One set were theories about the ease of finding ships in the North Sea; the other set were theories about the best way of attacking ships.

The results were fairly clear cut. The enemy could be found by mounting long-range reconnaissance patrols. His ships should then be attacked by high level bombing rather than by low level torpedo runs. Bomber development would therefore concentrate on early destruction of the German surface Navy by high level attacks, using heavily armed aircraft flying in Box formations. The new aircraft would be built to give a uniform range, to go beyond the mouths of the Scheidt and from east coast aerodromes to command the ports of the North Sea from Bergen southward and eastward to Kiel. The concept of the Box of fire provided by the day bomber formation seemed eminently reasonable, except to fighter pilots, and was taken up by the emerging Luftwaffe and by the US Army Air Force.

The problem for fighter defence of the country depended on the development of some kind of early warning system, and on an easier method of communicating than tapping a key and listening to dots. The course of the invention of radar and the setting up of the system, like the construction of the fast fighter controlled from the ground by means of radio-telephony, is well known. In these enterprises the British did no more than keep well up with the state of the art: when, after an accident the US Army grounded the Boeing P12, *Flight* remarked that it was meant as the American answer to the fast British fighters.[9] Americans, Germans, Russians, French and Japanese all moved at about the same rate towards the adoption of the very fast cantilever-wing multi-gun monoplane fighter; Americans, Germans and Japanese all invented radar in the laboratories. It was only the British who moved beyond the laboratories into the real world of detection and integrated it with a system of fighter control. And it was the British who had fighters in real organized squadrons. Everyone else dreamed about it as with the jet engine: but as with the jet engine, it was the British who delivered.

It is worth noting that all figures quoted for the performance of aircraft must be treated with some care. In particular range figures given by the manufacturers for bombers may be figures for full load, half load or empty, and we are seldom told which is being used. The manufacturer trying to sell his aeroplanes, or the Air Force spokesman trying to deceive the outside world, including his own taxpayers, tends to provide figures which are adapted to his particular purpose. Even if the design figures given are reliable for new aircraft flown at the factory by the maker's test pilot, the in-service performance will be affected by a number of factors which may be seen as the responsibility of the squadron system: principally, the skill of the pilot; the load or overload; the amount of new equipment or armament added to the original design and usually hung on the outside on the Christmas Tree principle, thus playing havoc with the aerodynamics; the same effect obtained from battle damage unrepaired or badly repaired; and the state of tune of the engine and even the smoothness of the paint finish. Nevertheless, a large-scale change in the range figures of new aircraft ordered and a new care for uniformity throughout the bomber force must mean a new strategic and tactical approach.

Similarly, to produce radar echoes and to claim long ranges of detection was one thing. What really mattered was the production of radar sets and their installation in strategic positions, and following this the development of a method of sending information to where it could be read by a commander and the orders conveyed to the aeroplanes. The large plotting table operated by non-technicians was borrowed from the anti-aircraft gunnery system; the girls who pushed the plaques around and even the controllers did not need to know how the information was gathered. But the planning of the telephone wiring inside the controller units so that each plotter got only limited information from the filter room and would not be tempted to make interpretations outside it, and the training of plotters not to exceed their briefs, in fact the whole paraphernalia of making a communications system and working it, was new ground. Inventing radar was easy: any genius could do it. But setting up a radar chain to work with a large body of fighters was something only the British had the patience to do. It needed someone working patiently and without the spectacular flamboyance of a Sykes, someone with the intense application of a Trenchard. But who was it?

There are a number of candidates, a group of, by then, very senior officers who had been chosen by Trenchard to follow him as Chiefs of the Air Staff. After the Salmond brothers came in 1933 another almost unknown figure, Sir Edward Ellington. He insisted that war could not happen then, in 1933, because his expansion programme would not be complete till 1942. Politicians took him to be foolishly complacent and so do historians: more subtle than they could imagine, he was saying that the Air Force would not and could not be hurried into any sudden panic measures. His first expansion scheme, the one that went to the Cabinet in 1934, planned for the expansion of the service to a force of eighty-four squadrons based in the United Kingdom by March 1939, to include forty-one bomber and twenty-eight fighter squadrons for home defence (Table 12).

Ellington's successor was Newall, the man who had been sent by Trenchard, originally, to set up the Independent Air Force. He had come from being Commander-in-Chief Iraq to be Air

Member (Member, that is, of the Air Council) for Supply and Organization in 1935. He became Chief of the Air Staff in 1938 and was thus still in post when the war broke out. But his first appointment to the Air Council marked a change in organization. He took over part of what had been the bailiwick of the old Air Member for Supply and Research: and this had been, since 1931, Air Marshal Hugh Dowding. Dowding remained on the Air Council as, now, Air Member for Research and Development, until his term of office expired, and then he took over Fighter Command. He was due to leave this command in 1939, but his successor, Sir Christopher Courteney, was injured in an air crash, and Dowding stayed on to fight his battle.

Thus it was Dowding who provided continuity for the beginning of re-armament. It was, of course, Dowding who was the moving force behind the introduction of radar and the making of the Control and Reporting system, not because he had any great emotional or intellectual commitment, but simply because it was by the structure of the organization his job and nobody else's. The accident which ensured that he remained after his allotted span to command in 1940 is one of the few recorded instances of justice being done by blind providence to men of worth.

During this period, there was a succession of international crises and of demands from politicians for more and more aeroplanes, followed by new plans for expansion to meet the crisis which had just passed. There was some official attempt to talk as if the major plans were realistic. The plans usually accepted as official were Scheme 'A', announced in July 1934 for completion in March 1939: Scheme 'C' announced in May 1935, calling for a force of sixty-eight bomber squadrons and thirty-eight fighter squadrons out of a total of 123 squadrons to be completed by March 1937, and usually claimed to have been completed on time: Scheme 'F' announced in February 1936 for completion in March 1939 calling for 124 squadrons with sixty-eight bomber and twenty-four fighter; and Scheme 'M' announced in November 1938, for completion in March 1939, with fifty-seven bomber squadrons and forty fighter squadrons out of 124.

The sequence is well described by a multitude of historians who deal with written documents and reports of speeches made for publication and take them seriously. There is no evidence that any of them were implemented except for Plan A. But there is no need for historians to believe, like the politicians of the time, that the making of an air force needs no more than the production of a large number of aeroplanes and that this process can be turned on or off like a tap. The real job of the historian is to look at the facts and then to use the documents to illuminate the facts. The following chapters represent an attempt to provide the facts which historians over the last generation ought to have started from in the beginning.

Notes and References

1. Peter Ustinov, *Dear Me*, Heinemann, 1979. Gorlitz, *The German General Staff*, p. 327. M. Cooper, *The German Army*, 1931–45, Macdonald and Jane, 1970, p. 69.
2. Folklore of the International Brigade: various personal communications.
3. E. Shinwell, *I've Lived Through It All*, Gollancz, 1973, p. 127.
4. H. Pownall, *Chief of Staff* (ed. Bond), Leo Cooper, 1972, Vol. I, p. 77.
5. See e.g. R.P. Shay, *British Rearmament in the 1930s*, Princeton University Press, 1977, pp. 38 et seq. As the author is writing for an American readership, he takes the trouble to explain many terms unfamiliar outside the British establishment of the period, and therefore comes high on the essential reading list.
6. B. Collier, *Leader of the Few*, Jarrolds, 1957, pp. 106 seq.
7. The most thorough coverage of the Air Exercises, written for a knowledgeable readership, is in *Flight* for 1933.
8. W.Cdr. J.L. McLean, 'The RAF Training Year', *RUSI Journal* 1935, pp. 50 seq.
9. *Flight*, 21 March 1934.

Chapter 10

Rebuilding

Priorities

Once the realities of defence in terms of available aircraft and pilots, and therefore of sorties available, had been stated, the questions would come, how do we improve the situation and how long will it take. The first answer was easy to give: the shape of an air force to defend the country was given in plan 'A'. But why could that not be ready till March 1939?

The time constraints were three-fold. The supply of aeroplanes was not one of them. There was no knowing now, in 1933, what types of aeroplanes would be needed in 1939, but preparations could be started to produce aeroplanes in general when the time came. It was important to have aeroplanes, and a lot of them, but they must be of the right types. There would be no question of going to war in another six years with a large stock of Wapitis built up carefully a few at a time. The aeroplanes could wait till the service was ready for them. In the event, it was Lord Weir, as what the Americans would call a dollar-a-year man, who came back into the Defence system to organize all that. That is another story.

There were three more important constraints on the rate of growth. It took two years to train a pilot to the state where he could take his place in the line of battle. It took three years to train a tradesman to where he could be wholly responsible, without supervision, for the maintenance of a military aeroplane. And it took four years and more to decide on a new airfield, and where to put it, to get the land, and then to erect the permanent buildings.

The Stations

The airfields built in the great expansion of the late 1930s are still there, indelible and independent evidence of activity. Each cost about the same as a cruiser, that is about £750,000, and it is likely that the Admiralty would have preferred that thirty cruisers were built instead. Dating evidence is harder to get, short of excavation: but for each one the first sign is a mention in the *Air Estimates*. In some cases the airfield when it first appears has its final name. Sometimes it is merely referred to as, say, 'Hampshire, a station', leaving us to wait for the following year for the name. A little care and local knowledge allows us to work out when an airfield was first proposed.

The location of the airfields was dictated by the state of the art of Air Traffic Control at the time. Mass take offs, as seen on the newsreels of 1940, were easy enough, but massed landings were not possible. Once a squadron was in the air, in 1933, the station couldn't know where its aeroplanes had got to, or ask them where they were. The first news of their return might be a garbled Morse message but more frequently was a visual sighting of an aeroplane entering the four-leg circuit, and then another close behind. It would take about a quarter of an hour to land a twelve-aeroplane squadron. Even if the land allowed, it was not possible to build airfields so that their circuits overlapped. Any package tour passenger, coming in towards Stanstead or Luton, crossing the coast at 30,000 feet, can see on a clear day the entire area of the Battle of Britain, and realize how crowded the air was that September. The Air Staff of 1933, allowing for the day bomber units for coastal defence, could only allot at most two hundred fighters for the 12 Group area. At the beginning of 1939, 12 Group Commander had about as many squadrons and aeroplanes as he or anyone else could handle in his air space. This calculation was radically altered when the capabilities of radar and the radio telephone could be appreciated. But since the airfields took the longest to establish, it is worth looking at them before examining the other development of the 1930s.

A permanent, or even semi-permanent location used by the RAF is referred to as a station, a term derived from Army practice, especially in India. Some stations are not airfields, and are therefore referred to as non-flying stations. These are such places as the Packing Depot, which was for so many years at Ascot, and then at Sealand. Stations which have, or really are, airfields are referred to as flying stations. Those which house squadrons are called operational stations: other airfields are not operational but house units which are essentially concerned with flying, such as flying training schools.

A completed permanent flying station of the pre-1939 expansion pattern incorporates certain essentials. There must be an airfield, which in the 1930s would be a grass field without concrete runways. Fronting on the airfield are normally a number of hangars, massive structures with sliding doors to give an entrance of ninety-eight feet wide, half as high, and the size inside, approximately, of a soccer field. It is normal to find one aircraft storage hangar for each squadron which the planners of 1933 saw the station as housing, and another fronting on the airfield for second line servicing. First line servicing, including fuelling, tyre checks and all the other work of what is, essentially, a filling station, is carried out on the apron. The men here work in the open air in all weathers, and are the last to see the aircraft before it takes off and therefore the last to be able to stop it if there is anything visibly wrong.

Along the sides of each hangar are permanent lean-to structures subdivided into smaller rooms, which are where the squadron, in terms of men, lives in working hours. There are workshops, stores, flight offices and crew rooms, the main function of the latter being to house arrangements for the constant supply of a liquid known as NATO standard, composed of hot water, NAAFI coffee powder, sugar and condensed milk in ill-judged proportions. In the 1930s one must presume that the powdered coffee was replaced by tea, or by liquid coffee essence out of a bottle. However unpalatable, it is often the last sustenance taken by brave men who flew off and never came back.

A stage further back from the airfield is usually another hangar which is used for major repairs which do not yet require the machine to be sent back to the makers. Associated with this is normally a Safety Equipment store which houses the parachutes and dinghies. The parachute began to come into use in the RAF in the early 1920s. The main reason for delay in equipping every pilot was the inability of the Irving Company in America, whose product seemed to be the most reliable on offer, to produce enough in a hurry to satisfy virtually every air force in the world. Sir Frederick Sykes, MP, continually asked penetrating questions about the rate of supply.[1] The date of 'adoption' of the parachute, so frequently given as 1925, appears to mark the completion of re-equipping. The memory of Maurice Baring was sufficiently catered for by a large motor transport detachment with its necessary garages and workshops. The road system of the station does not normally include footpaths, which increases the excitement of airmen in their well camouflaged uniforms walking about at night. In some obscure place is a chapel, usually dedicated to St Michael the Archangel.

The Army ancestry of the service was remembered in a large parade ground, or square. Later radar stations, built in concealed sites in great secrecy, could be recognized from the air by the presence of a Square. Around the Square were usually found the airmens' quarters in barrack blocks built each on an H-shaped plan, on two or three storeys. The long legs of the H were taken up by the barrack rooms, while the toilets and ablutions were in the cross-bar. At the junction of the barrack room and the ablutions was a limited number of small single rooms. These were normally the perquisites of corporals, promotion bringing with it this modest ration of privacy. Nearby was also an Airmen's Mess, and sometimes in a separate building a corporals' club.

Further from the airfield was the Sergeants' Mess. Here the sergeants and warrant officers slept and ate, and had a bar which relieved them from using the Canteen (or later, the NAAFI.) In any case, existence in this entirely masculine society was comfortless and barren. In 1938, the Air Ministry decreed that airmen aged over twenty-one who had accommodation might claim as of right permission to live off the station, and draw lodging allowance.

Furthest of all from the airfield was the Officers' Mess. The standard station was designed by Sir Edwin Lutyens, and erected everywhere regardless of the lie of the land. Expert advice was taken on the siting of the station so as to fit into the landscape, but the results of this advice are not always immediately apparent. No landscape could quite hide the huge hangars, and a station can be spotted from afar, even from the air, by looking for the water tower overshadowing all. But the Officers' Mess, often across the public road from the rest of the station, showed the signature of its architect plainly in its structure.

There are two main types of Lutyens messes, two-storey and three-storey. They differ also in their internal arrangements. It is tempting to guess that the types were for two-squadron and for three-squadron stations respectively, but that theory does not stand up to facts and observation: possibly minds were changed after ordering. In any case, the mess had a large dining room, which was itself the mess by derivation, and usually four sitting rooms, or anterooms. The ceilings are high and the cost of heating therefore enormous, but that did not matter in 1933. Behind the imposing front and above the large public rooms were ranges of bedrooms. The impression which these buildings attempt to give and the way of life they encourage, is something between the country house and the large hotel.[2]

The exterior of these messes is reminiscent of the monumental appearance of even Lutyens' smaller country houses. They are meant to encourage a dignified and sedate manner of life. It should be remembered that by the standards of today the officers of the 1930s were in manners and behaviour rather stuffy: this stuffiness even underlay the rake-hell image cultivated by the short-service officers. The officers of the pre-1914 period were by our standards incredibly stuffy. The routine of mess life was based on that of an Army regiment: perhaps not on one of the best regiments, but certainly of an equal degree of formality. As in a Victorian regiment, all officers dined in mess on most evenings of the week. Dinner was a parade, and attendance compulsory. Mess dress, that is a military version of the civilian full evening dress, with a short 'shell' jacket and tight trousers, 'overalls', were worn with white shirts and bow ties. Most stations now hold a dining-in night once a month, and combine it with the old Guest Night, then the one night in a week on which members might bring in guests to dinner. Guest nights were a very formal occasion, when boiled shirts and white ties were worn. Civilian guests wore full evening dress, as contrasted with the dinner jackets now thought sufficient.

Most of the ceremonial trappings are now symbolic but were

originally highly functional in their Army setting. Mess dress was originally a comfortable form of uniform, and it was worn in place of the normal working uniform to show that the routine of the day was over. The RAF's use in the 1950s of an interim mess dress, to relieve National service officers of the cost of buying full mess dress, consisted of the normal uniform worn with soft dress shirt and bow tie, and missed this point completely. The passing of the port was a process devised when port was the normal drink of gentlemen, to slow down the rate of drinking and therefore the onset of drunkenness. The ban on politics, religion or women as topics of conversation was an expedient to prevent quarrelling among slightly drunk officers after the meal. These formalities were already dying out even in moneyed civilian society in the 1930s. But the large rooms and high ceilings of a Lutyens mess made such a ceremonial daily life seem natural.

Mess life, whether in Officers' or 'Sergeants' Messes, was a single sex activity. Ladies might be admitted as guests on great occasions, as a Station Ball or a special Ladies' Night, but women had no place in the military society. The appearance of ladies as officers of a kind, with recognizable places in the military operating system, posed a variety of social problems of status and entitlement to mess managers and mess committees. The mess is governed by a mess committee, whose president is usually the most senior officer on the station below the Station Commander. While in a warship the Captain is not a member of the Wardroom, the Station Commander is a member of the Officers' Mess.

The 'Lutyens-Stations', as we must call them, proved more durable and useful pieces of military equipment than cruisers. Most of them remain today, and while many are still in RAF hands, others have been re-used as motor racing tracks, as open prisons, as hospitals for the USAF, as hostels for Vietnamese boat-people, as tomato farms, and even, in a final degradation, as Army barracks. It is often remarked that the Army value these bases so highly that the first action of the soldiers on gaining vacant possession is to dig up the runways in case the RAF might want them back. Lutyens' son, in a memoir, points out that an architect was then paid a standard 10 per cent of the value of the work, out of which he had many overheads, and the construction of so many identical major buildings must have been profitable to the architects' office. Sir Edwin Lutyens died in 1949.

A permanent station when built must have had a massive impact on the local scene. They were, of course, sited in the countryside, usually fairly far from large towns, and overshadowing the villages to which they were neighbours. They were built by the large building contracting firms of the time. Construction was not merely a matter of flattening land and putting up the buildings. Roads and even railway sidings had to be built, electricity and water had to be laid on in sufficient quantities, and drainage provided. The latter posed a problem to the localities where a village of perhaps a thousand souls found as many airmen again arriving on the newly built station. It is credibly reported today that the only limit on the size of Cranwell is that posed by the capacity of the local sewage system to deal with the human waste produced.

It may be noted that the practice in Britain is for a military station to be named after the civil parish in which the station headquarters is located. This accounts for the situation at, for example, the single site within one perimeter fence which accommodates the Royal Military Academy at Sandhurst and the Army Staff College at Camberley. In the case of airfields, this explains why the vagaries of parish boundaries mean that a station main gate is often quite far from the village which gave it its name, and near to or even within a village with quite a different name. It all helps to make map reading more interesting for visitors.

It was not, during the process of expansion, considered necessary always to wait for the completion of the permanent brick buildings before taking the station over and operating it. As soon as the airfield itself was cleared and the mains services put in, the station could be set up in temporary huts.

Station Functions

The general picture may be filled out by examining one station in May 1938, in this case RAF Upwood, chosen because it was while working here that the author first conceived this book. The Station Commander at this time was a wing commander. The *List* shows that he had an adjutant and an assistant adjutant, both employed as civilians, although they were retired RAF officers with ranks in the Reserve of Air Force Officers. (An adjutant in the British services is an administrative officer, unlike in the German and other foreign services where the word is used to mean what we would call an aide de camp, and should be so translated.) The Station had two equipment officers, who were both civilians and both retired Army officers. There was a regular flight lieutenant of the Accountant Branch, with under him a retired major. The medical officer was also a regular flight lieutenant. There was an Education Officer Grade III, who was a permanent career civil servant and therefore nearer in standing to the other three regular officers than to the retired and re-employed civilians. The whole complement of the Station was thus three regular RAF officers and six civilians who were of officer status, and therefore affiliated members of the Officers' Mess. There were also no fewer than five warrant officers on the Station strength.

The Station housed two day bomber squadrons. Both were to become Group Pool squadrons when this system came in during January 1939, but it was not yet effective. 52B had been re-formed in January 1937 with Audax and Hinds, and later in the year moved to Upwood. There over the winter of 1937–8 it was re-armed with Battles. 63B had been reformed at Upwood in February 1937 from 'B' Flight of 12B, with Audax and Hinds, and re-armed with Battles, although it also had Ansons on its strength 'for training'. This does not look like an orderly or well planned succession.

By the *List* of May 1937, 52B had on strength one squadron leader, one flight lieutenant, and one flying officer, besides eleven pilot officers and five NCO pilots. 63B also had a squadron leader, a flight lieutenant, and a flying officer. There were ten pilot officers, two of whom had the 'sn' annotation, and another of whom was 'first commissioned 31/5/17' according to a note in the *List*. There was one pilot officer RAFO, one pilot officer RAFVR, and five NCO pilots.

The total complement of the Station, and of its squadrons, that is of the Officers' Mess, was thus thirty-one officers and seven civilians of officer status. Of the officers, seven might have been in the service more than three years – the doctor may well have been freshly commissioned in what was the beginners' rank for his branch. The wing commander and perhaps the doctor may have been married, and the civilians were probably for the most part married and elderly, except for the education officer. He, like the other officers, was almost certainly a bachelor, and lived in the Mess.

There would have been something like 700 airmen: it is almost an article of faith that the most difficult thing ever to discover is how many airmen there are on a station, with all the complications of attachment and detachment and leave and sickness and desertion, each in a dozen different varieties and degrees of permanence. When the Station opened, the main services, the drains and the water and the electricity, were laid on, but most of the permanent buildings were not yet completed. The men, and probably the officers, lived for the most part in huts scattered over what was a confused muddy building site infested with workmen and loosely controlled by the Department of Works and Bricks.

Upwood was a convenient place to be, socially, as compared with other new Stations, Little Rissington, say, or Driffield. There was a pub ten minutes walk from the main gate, and it took only half an hour to walk into the town, Ramsey, where there was a cinema, and shops. Even in those days it was unusual for a junior officer to own a car, or even a motor cycle, and for an airman it was almost unknown – T.E. Lawrence, with his Brough, the relic of earlier and richer days as one would expect in the gentleman ranker, was a rare exception.

The Station was therefore a closed community, or rather two closed communities, one of officers and one of airmen, with the sergeant pilots occupying a slightly ambiguous position. In this and in many other closed groups, virtually isolated from civilian life, during the time from autumn 1937 to summer 1939, the ethos of a new Air Force was hammered out, with a small legacy from the old short-service days; and with the ethos came a vocabulary, a new language of understatement and technical reference which in the 1940s everybody was to learn. These men in their isolated stations were the men who fought the Battle of France and the Battle of Britain.

Administration of this Station, with the large numbers of men, and junior officers, still unused to service life, and all set about with the contractors' men always under their feet, must have been a nightmare for the Station Commander and his staff, although their main guide book was clear; this was *The King's Regulations for the Royal Air Force*, a version of *The King's Regulations for the Army*, with only the names changed where necessary to protect the guilty. Everything must have revolved around the ability and experience and good will of the five warrant officers. Probably things did not always run smoothly, but the supply of geniuses in the country has always been limited.

Reliable commentators of the time are agreed that the squadron commander did not get into the air very often, nor did his deputy. 'Command of the squadron in the air devolved on the most senior of the Pilot Officers, usually a man with less than two years service.'[3] The squadron commanders were submerged in paper work. What were they doing?

The two squadrons were re-arming with Battles. Re-arming a squadron, or a station, is not merely a matter of flying out the old aeroplanes, and flying in the new ones. Before the new aeroplanes come in, you must re-man the flights with men who are qualified to service them and maintain them. The right number of tradesmen must be withdrawn from their posts scattered all over the service, anything up to a year ahead, and sent on courses within the service or at the makers' factories. These men would be the NCOs supervising the servicing crews on their new squadron, or station. The basic training courses for apprentices at Halton and at Cranwell would have to be rearranged two years or more ahead so that trainees could be directed in precisely the right numbers, allowing for training wastage (in other words, failures) into the right specialist streams and classes where they would be taught by properly qualified and trained instructors. The same would apply, over a shorter time scale, to the training of lower skilled men at Manston, and to the provision of the right number of cooks and clerks. In a word, to ensure that a Battle squadron could be formed or armed at Upwood in 1938, the Manning Plan for 1934 would have to take account of it. Without computers.

But the aeroplanes were not the only hardware to be supplied. The airframes and engines would not operate unless a whole world of spares were available in advance, all delivered and unpacked, and put away in the right order. Besides the actual spares, a variety of custom-built workbenches and jigs and machine tools would have to be brought in and set up. Everything would have to be unpacked and checked and signed for, and sent back with complaints if it were not according to the specification. That meant a sea of paperwork, and though the apprentice clerks had been trained to deal with it, and the boy clerks before them, there were not enough of them. Nor were there ever enough typewriters, and photocopiers had not been invented. Everything in the end fell on the squadron commander. Trenchard's Air Force, which had no specialist administrators, had not been erected to deal with the problems of a rapidly expanding service and a massive dilution of all skills and trades.

There are two points here to be noted. One is that when we read that a squadron was re-armed with a given new type of aeroplane on such a date, this does not mean that the squadron could operate that type on that date. The aeroplanes would not all arrive at once, but would dribble in over weeks, and never on the dates promised. The same, of course, applied to the servicing manpower and the servicing machinery, and even to the pilots. And once the pilots had come in they would have to learn, or be taught, how to fly the new aeroplane and carry out the new task. Teaching would just not be possible if the very few senior pilots were too busy with the paper work to fly: which they usually were. Planning in 1933 was not perfect.

The other point is that when the German attack on the fighter airfields in 1940 forced squadrons away from their bases and into temporary airfields, either to relief landing grounds or to flying clubs, they gained a massive victory. It was not merely a matter of causing a little personal or social inconvenience to the British pilots. Once the squadrons were driven away from their servicing installations, from their stocks of spares and their benches and jigs and machine tools, they were only a short time away from ceasing to operate altogether. It was the equivalent of closing the Royal Dockyards to the Grand Fleet.

The Stations

How many stations were there? There were to begin with in April 1933, forty-three flying stations in the United Kingdom. Of these, three were flying boat stations, two were 'bases' for naval squadrons ashore, six were civil airfields which housed auxiliary or cadre squadrons only, three were flying training schools, one was the Central Flying School, and one was the RAF College. This leaves twenty-seven stations housing operational squadrons. Seven were fighter stations, and of these three had only one squadron each. Four had Army co-operation squadrons, although one of these was Farnborough which had other business for its airfield. Martlesham Heath had nominally two bomber squadrons but was not, as we have seen, really a bomber station, while Donnibristle had the one torpedo-bomber squadron. This left fourteen operational bomber stations, of which Boscombe Down and Bicester had one squadron each, and Worthy Down and Upper Heyford had three squadrons each (Table 13).

The course of expansion can be followed from the annual *Air Estimates*. The amount of money to be spent on each station, either on re-building of old stations or on the creation of new ones, is laid out annually. The *Estimates* are not always clear on whether the money is for a new station or for an old one, but a little work brings out the following picture of expansion.

In the *Estimates* of 1934 one new operational station is announced, Mildenhall. In the following year, in the *Estimates* for 1935, we see two new flying training schools at Ternhill and Hullavington, and eight operational airfields, approved for construction. In the 1936 *Estimates*, we find two flying training schools at South Cerney and Great Rissington (later Little Rissington), a Research Establishment at Bawdsey, two armament training camps at Peaches and West Freugh, and twelve new operational stations. In 1937, we find ten new stations, of which five are operational flying stations, the others being a hospital, two armament training camps, an equipment depot, and a wireless telephony station in the Scilly Islands. In the *Estimates* for 1938, there are nine new stations, including four armament training camps, two schools of technical training, two equipment units and an aircraft depot, but no operational flying stations except for enlargement of Detling to take one Auxiliary squadron. Real estate expansion had stopped till the war started. The armament training camps are usually specifically described as 'hutted'.

One of the more interesting new stations is RAF Cosford. Halton was about as bad a site for an apprentice school as could be imagined. A main road runs through the site, with the domestic buildings on one side of the road and the classrooms, workshops and airfield on the other, and classrooms and dining halls as far apart as can be inconveniently managed. The congestion at mealtimes is for connoisseurs to appreciate.

A new apprentice school was to be built at Cosford, on a unitary site conveniently close to a railway station. To avoid the problems of Halton, the design called for one building, a squat brick tower of forbidding aspect.[4] The classrooms and the workshops were on the ground floor. The apprentices' accommodation, their dormitories, dining halls and leisure centres, was all on the upper floors. Those who go to Cosford to watch the indoor gymnastics may admire it from afar. A similar layout was used at Shrivenham, originally designed to house a regiment of artillery. Here are four brick blocks, each meant to accommodate one troop, with the guns and vehicles on the ground floor and the gunners living on the floors above. Like the airfields of the period, the design has proved adaptable.

The interpretation is fairly simple. By 1937, practically the whole of the war plan was complete as far as operational aerodromes were concerned. All that were planned for use had been started. There were in all fifty-three operational flying stations either completed or begun, and available on the outbreak of war in 1939 (Table 14).

Organization

But the layout of an air force needs more than airfields. The original airfields of 1920 had been chosen not quite at random from the airfields remaining from the war. There was a circle of airfields of differing degrees of sophistication around London which were obviously worth retaining, and several stations on the coast which were needed for the anchoring of flying boats or the housing of carrier aircraft when not at sea, and for the one or two torpedo-bomber squadrons. The bomber squadrons had to be based somewhere, and as we have seen they were kept in the Army's manoeuvre area on and around Salisbury Plain. The original organization had grouped the units into Areas: a Coastal Area, an Inland Area, and a Fighting Area.

By 1932, the Areas had settled down, and for the most part showed a functional rather than a regional organization, although the Area nomenclature was retained. Coastal Area controlled all its stations and the squadrons embarked in the carriers in home waters directly from a headquarters now at Lee-on-Solent. Besides the squadrons and schools, the units included a Marine Aircraft Experimental Establishment and a Flying Boat Development flight at Felixstowe, a Torpedo Section and workshop at Gosport, and a floating dock at Pembroke Dock. The Commander-in-Chief was Air Vice-Marshal R.H. Clark-Hall.

Inland Area had its headquarters at Bentley Priory, and was headed by Air Vice-Marshal A.E. Borton. It controlled directly the Air Ministry Wireless Section, in its gas-proof room at the top of Adastral House, the Armament Training Camps and the RAF Band.

But it had three Groups. 21 Group, HQ at Digby, was principally concerned with the four Stores Depots and the Home Aircraft Depot at Henlow, the Stores School and the Medical Stores, but it also had on its books the Depot and Hospital at Uxbridge, the Record Office, and the Experimental Depot at Martlesham Heath. Obviously it was the beginnings of the later Maintenance Command.

23 Group had its HQ at Grantham. It controlled the three flying training schools and CFS, and the School of Technical Training (Men) at Manston. To this training empire, the forerunner of Training Command, was appended, probably for geographical convenience, the Packing Depot at Sealand, which would seem functionally to belong more to 21 Group.

But 22 Group seems even more awkwardly placed in this Area. The other two Groups were basically non-combatant organizations. 22 Group commanded the School of Army Co-operation, the School of Photography at Farnborough and all the Army co-operation squadrons in the UK. It also controlled the Balloon School at Larkhill. It was, in fact, the last stand of the old Military Wing.

The main functions which the Naval Wing had developed, fighter defence of London and long range bombing, were now the business of a Command named Air Defence of Great Britain (ADGB) with headquarters at Uxbridge. Its primacy was shown by its being commanded by an Air Marshal, at this period Sir Geoffrey Salmond. It controlled no units direct, but had three main formations. 1 Group, with headquarters in Tavistock Place, controlled all the cadre and Auxiliary squadrons, and also the Central Medical Establishment and the Inspectorate of Recruiting, presumably because they shared the building with Group HQ. It has no obvious descendants, unlike the other two formations in ADGB.

The Wessex Bombing Area had its headquarters at Abingdon. It controlled directly all the bomber squadrons and their stations. The RAF Staff College shared Andover with the HQ, and so was conveniently brought under its command. Oxford University Air Squadron was in the area, and therefore also brought under this headquarters. Its obvious descendant is Bomber Command.

The ancestor of Fighter Command was Fighting Area, with headquarters also at Uxbridge. This controlled for convenience the Anti-Aircraft Co-operation Flight and Cambridge University Air Squadron, and also directly all the fighter airfields and squadrons.

This organization, now essentially functional, was the basis for the development of a much larger Air Force with much more definite operational tasks. The first step came in September 1933. A new headquarters was set up for Central Area. Western Area was now to replace Wessex Bombing Area and to control the night bombing squadrons. Central Area was to control the day bombers.

The ensuing development was consistent. The new stations were allotted according to their function to what became Training and Coastal Commands in 1936, Bomber Command and Fighter Command in 1937, Maintenance Command in 1938. The wartime system of Groups, completed only in 1939, was hardly in place before the war started (Table 17).

The Empire

There were other places besides Britain, and other squadrons and airfields. The course of expansion in these areas was leisurely and rational. It may be argued that there was some kind of struggle between an Eastern and a Western faction in the Air Council, but the material does not show much evidence of a change of direction.[5]

There was, after all, not much of a struggle possible. The air forces abroad were basically what the Army required for its support, and what the Navy asked for. It was evident that if any one element of the overseas Commonwealth were overwhelmed by an enemy, then a war could be waged for its recovery from the safe base of the United Kingdom. However, if the United Kingdom were overwhelmed, then the Commonwealth must also go down. The first efforts must therefore be devoted towards the protection of the United Kingdom. The build-up of air power overseas could wait till that was accomplished, and till the changing shape of politics showed where the aeroplanes were needed (Table 9).

Uniforms
In the depths of the Empire, the airmen wore the same uniform as the Army, with baggy shorts and pith helmets resembling extinguishers. But at home, things were changing.

While the form and routine of the station before the expansion showed very clearly that it was hardly modified from the Army pattern, the uniform was also clearly the same, except for its colour. This dress had been developed for use in the Boer War, and altered a little before the First World War. Airmen up to the rank of corporal wore choker-collared tunics of a heavy hairy wool. Sergeants and flight sergeants wore tunics like those of the present day, with collars and ties, a relic, it was said, of the privileges of the petty officers of the RNAS. Officers and warrant officers wore tunics of this pattern, usually made to measure rather than issued from stores, of a smoother more comfortable cloth, and with white shirts and starched stiff collars. Everybody wore trousers, 'slacks', for work or informal duty occasions, as what the Victorian military tailor would have called 'undress'. On formal occasions, all ranks wore baggy knee-breeches. Other ranks wore heavy ammunition boots with puttees: officers might wear puttees, but are more often seen in photographs wearing either knitted woollen golfing stockings, or knee-length field boots of leather. They had batmen to do the polishing.

The reforms of 1 April 1936 were intended to modernize and liberalize some aspects of the service. From this date, all ranks were to wear the open tunic with collars and ties: the shirts and soft or semi-stiff collars in all cases were to be blue. Everybody was now to wear trousers even on formal parades. The breeches, and the puttees which had caused so much trouble to Maurice Baring, were abolished.

Up to this time, the only headwear had been the peaked cap. Now, however, the soft side hat introduced by Sykes was brought back for informal occasions, and it was optimistically hoped that it would also serve as a flying helmet. At least it was easier to stow.

RAF Officers had never worn Sam Brownes to keep their scabbards still while they drew their swords. The sword is an article of uniform, and a stock of swords with belts and scabbards is kept at a number of strategically placed stations for loan against signature to officers who need them for great parades or weddings. Maintenance Command presumably housed a director of sword play.

There is an almost theoretical full dress for the RAF, which may be seen in some portraits, although it is debatable whether the officers actually bought it rather than hired it for the sittings: it is more or less the Royal Engineers full dress in a different colour, Hussar busby and all. The Central Band still wear a version, complete with busbies.

The RAF allowed the optional wearing of a moustache, like the Army who had made its growth compulsory, whatever the capacity of the wearer, until 1916. Both services drew the line at beards. Like the naval beard, worn at that time mostly by aviators and submariners, the RAF moustache became the mark of a *corps d'elite*. It was the only way in which these men could express their individuality.

Notes and References

1. Sir Frederick's Parliamentary Questions were all reported in *Flight*, and provide a fertile source of unexpected information.
2. J. Nesbitt-Dufort, *The Black Lysander*, Jarrolds, 1973, pp. 35 seq.
3. B. Embry, *Mission Completed*, Methuen, 1957.
4. *Flight*, 18 June 1938.
5. See Malcolm Smith, *British Air Strategy between the Wars*, OUP, 1984. No reference is made to actual numbers as shown in the *Lists*.

Chapter 11

New Foundations

The Beginning

In what is usually called his 'Parity' speech, Baldwin said that it was all very well to call for more aeroplanes. But aeroplanes were of no use unless you had men to fly them and aerodromes to land them on. He was quite right as far as it went, but he had left out the bulk of what was needed for an air force.

An air force needs pilots. It obviously needs men to maintain the aeroplanes, to fuel them and re-arm them and turn them round, to carry out routine servicing, to repair damage caused by accidents and by battle. But it also needs cooks and clerks, electricians and telephonists, storemen, lorry drivers, nurses and sick bay attendants, its own police and drill instructors and firemen, general labourers to fetch and carry and clean things and places . . . oh yes, and the band.

Trenchard's original quarrel with Rothermere was about these men, the need to have them, and to have their training and the places where they worked under the control of the Air Staff. Rothermere did not want to have them controlled by the Air Staff because he did not believe, and neither did anybody else who mattered at the time, that the Air Force would last. But it did last, and that was largely due to Trenchard's refusal to let it die and his willingness to stand up face to face with the generals and the admirals and the politicians and fight for it. The Air Staff knew it had to have these men for expansion, and went away to re-write the manning plan.

Manning

Academic historians and social scientists often imagine that an army, or even an air force or a navy, can be raised in a hurry by pulling in a large number of men off the street, the more the better. It does not work like that.

First of all you decide what you need an air force for and when: that is, you make out a Task Chart for the whole service for the date you expect to make war. By the time you finish the operation you may have to alter the Task Chart, just as you may have to alter aircraft specifications, but that is a detail and what management is about. When you have worked out exactly what units you need in the front line, you can ask what second line and training units and headquarters units you need to support them.

From that you can make a table – a Manning Plan – of how many airmen you need in what trades. For each trade – military speciality is perhaps a better term – the men need to be trained, and the length of the technical training plus the length of the initial military training tells us how long before a unit is opened you have to start training the men who will go into it. Thus you may have to recruit most of the aero-engine fitters three years before the unit opens, and most of the cooks six months. But some will have to be brought in and trained earlier so that you have experienced supervisors – junior NCOs – on the stations. Before you can train any of them you have to set up the training schools and find the instructors mainly from the skilled men already in the service. And these instructors will have to be replaced as they are withdrawn from their front line units . . . Remember that there may be as many as forty technical specializations involved, each one with a different training period, and a moment's reflection should convince you that the construction of a manning plan for five years in the future is a very complex operation in itself. And in 1933 nobody had invented, or even thought of, computers.

The Trenchard manning plan had been comparatively straightforward. There was no time pressure. The theory was that every individual aeroplane was looked after as in the days of Biggles by two craftsmen, a fitter (that is, an engine tradesman) and a rigger (airframe tradesman), both ex-apprentices trained at Halton. They were assisted by a small number of aircraft hands, or general labourers, who were unskilled men and might be employed anywhere else around the station as required. The original aim had been to have the skilled men accounting for about 60 per cent of the airmen. Allowing for about 2,000 apprentices entering each year, with perhaps a 20 per cent wastage in training, and coming into service, after 1923, for eight years active service, and a proportion of these being allowed to serve on for longer terms, it seems likely that in an Air Force of 20,000 men, this proportion was being approached by 1933.[1]

The other trades were filled by recruiting men with, as far as possible, formal civilian trade training attested to by documents. They were given initial military training at the Depot at Uxbridge, and then went for trade training at the Technical Training School (Men) at Manston. Sykes' plans had gone no further in sophistication than this. Since a great proportion of these adult entrants to unskilled or semi-skilled trades would be employed at second line non-flying units, like Stores Depots or Hospitals, it is likely that by 1933 the proportion of the men manning Operational Stations and Flying Training Schools who were skilled ex-apprentices was well over 60 per cent (Table 20).

However, it was obvious that if the service was to put eighty-four squadrons into the air in 1938, there was no chance of training 60 per cent of the manning requirement up to apprentice standards. The Air Staff therefore had already made two decisions before the beginning of 1933: one was to dilute the trades and the other was to use the existing apprentice-trained men sensibly.

The dilution was begun in 1933. It was announced that men were now to be recruited direct from civil life as what were then to be called 'Mates', that is Fitters' Mates, a recognized category of semi-skilled worker in civil life. They would be taken in through Manston and trained entirely in the service, but not to the level of the skilled Fitter. Hindsight sees the beginning of the later service stratification of Fitters and Mechanics.

The sensible use of trained skilled men was parallel. The old trade demarcation between fitters and riggers was to be abolished. From now on, Halton apprentices were to be trained as 'Fitters', capable of any aeroplane job whether on engines or on airframes. The apprentice would pass out with the category of fitter II, and go to a flying station. At his station, the fitter II would be employed on the flights under supervision, with systematic instruction and intensive conversion courses back at Halton to bring him up to the standard of a trade test, after which he would become a fitter I. The existing fitters and riggers of the old system would be sent back to Halton and brought up to fitter I standard on a formal course. The fitters I would be employed on the really difficult work in the workshop hangar. The main duty of the fitters on the flying station would be to supervise the mates, who would actually do the routine work on the aeroplane.

In the 1934 *Estimates* allowance is made for 933 apprentices at Halton and Cranwell in the trades of fitter II, twenty-nine as fitter armament, thirty-seven as fitter aeroengine and forty-eight as metal rigger. There were also 292 wireless operator/mechanics, and sixty-six instrument makers. In comparison, Manston provided for blacksmiths (a one-year course, with an average of six men under training), motor transport drivers (four months and sixty men), fabric workers (six months and eleven), fitters aeroengine (four months and sixteen). There were also to be an average of fourteen men converting from fitter motor transport to fitter-driver in six weeks, and 110 riggers converting in four months to metal rigger.

It had been an essential feature of the Sykes concept, carried on by Trenchard, that the ground tradesmen should be part of the squadron, or even a permanent part of the flight and attached to one particular aeroplane. Under the new system a large proportion of the skilled workers would belong to the station, not to a squadron, and would be employed in the station workshops, in the second row hangar away from the apron: the station would be operating one type of aircraft, using one bank of spares and benches and jigs and expertise. The men who did the aircraft turn-round and the minor and immediate repairs and simpler routine servicing were still attached to the squadron, if not to an individual aeroplane. Already there were signs of the future. Some stations were experimenting by the late 1930s with the 'garage' system where all the engineering work was done by one technical wing, irrespective of the squadron, but nobody really wanted that: the squadron concept, as a real society embracing all ranks and all trades, died hard. But the squadron as a unit was being replaced by the station. This was imposed by the growing complexity and physical size of the aeroplanes.

So the regular fitters I, out of Halton, were to supervise the mates and train on the fitters II. It made sense. But who were to supervise the fitters I?

The tradition we have seen. The ground tradesmen were supervised on each squadron by a pilot as a secondary duty. This was usually an officer who had been trained as an aeronautical engineer at the Home Aircraft Depot at Henlow, who therefore showed the 'e' annotation after his name in the *Air Force List*. If he had been a short-service officer on starting the Henlow course he now had a permanent commission like the Cranwell and University entrants. His formal contact with the men was exercised through the squadron sergeant major. (The title of sergeant major was abolished on 1 January 1933, and replaced by the term 'warrant officer': there were two classes of warrant officer, but these were merged in 1936.)

A solution was fairly obvious? If the unit were to be given a full-time officer to supervise these ground workers, someone who did not count himself as a pilot mis-employed and saw this supervision and not flying as his primary task, then unit technical work could be got into order and kept that way. The problem was to find these officers.

The problem of supervision was not a small one. The *Air Estimates* for 1937 allowed for a service of about 43,000 other ranks, together with about 7,500 apprentices and boy entrants in training. Over the next two years it was intended to bring in 21,000 adults for training. Of these, 5,900 were to be mates, 2,000 wireless operators, 1,300 fitters, 890 armourers, 570 cooks and butchers. And the largest single group were to be 7,500 aircraft hands, the unskilled general labourers who fetched, carried, cleaned, dug, pushed, and generally by the use of muscles, intelligence and good will made the whole system work. These men, entering before the spring of 1939 and in many cases not in productive service before that autumn, would be the last additions the RAF could definitely count on having before the expected beginning of the war.

The additions to the apprentice scheme must be noted. In 1934, the Ministry had announced the creation of a new class of Apprentice Clerks, to be recruited from holders of the School Certificate. The following year, they created a new type of entry, the Boy Entrants, chosen from those who had applied for apprentice training, but had failed to get in. They would be trained over two, not three, years, up to mechanic standard in the trades of armourer, wireless operator and photographer. As we have seen, by 1934, there were also apprentice instrument makers to do more delicate work than that of the hammer mechanics.

Somebody had to supervise these men. Someone had to watch them working, to inspect their work, to assess them, get them promoted, judge them when they were put on charges by their NCOs. Somebody had to do the progress chasing and make sure the complex system of forms and signatures was followed so that responsibility was accepted by every man for his own work, knowing that responsibility could be brought home to him. Somebody had to tie in the supply of spares and materials with the working rosters, and co-operate with the Stores Branch officers who might know the system of getting stuff but did not really know how it was to be used. The supervisors themselves would have to be supervised by a hierarchy of senior officers. And when necessary the supervisor should be able to enthuse his workers and lead them in emergency and in danger to long hours of unremitting and careful skilled labour. Supervision was a full-time job: the Air Force had come a long way from Biggles sitting outside the hangar watching *his* fitter and *his* rigger tune *his* personal aeroplane while he loaded his own belts.

Engineering Officers
The need for supervision had already dawned on junior GD officers, especially the pilots with the 'e' annotation who had been trained as engineers at Henlow. A semi-official journal had been started in 1930 as the *RAF Quarterly*. This served partly as a means of giving the service official information on what in the world was happening to it, and partly as a vehicle for expressing the views of usually junior officers whose names were not given but who passed under pseudonyms. The *Quarterly* continued to publish till the 1970s, when it was done away with on the grounds of economy. In October 1935, 'a Flight Lieutenant 'e' ' started a series of articles by different authors which continued into the following year.[2]

In this article, entitled *An Engineering Branch for the RAF*, the author called attention to the sad fate of 'young and carefree aviators' ordered to specialize. This is the theme that underlies everything. Officers had joined the service to fly; they enjoyed flying and resented having to do anything else. They rejected Trenchard's dictum that aviation is not a sport, it is war. In war, it is assumed, all Engineer Officers, that is officers 'e', will return to flying, and so will all NCO pilots who, having finished their original flying tour, are now working as supervisors in their trades.

This writer suggested that all would be well with the world if every GD officer now employed as an engineer should immediately return to flying, except for those who instead of going to Henlow had been sent on a university course. A new Engineer Branch should be formed entirely from University graduates with low medical standards. These men, after commissioning, should be sent for a year's attachment to industry and then employed at the Aircraft Depots: not at squadrons.

It was also pointed out that there were 130 scientists employed at Farnborough. These could be replaced by officers from the new Engineering Branch. This is a constant theme of the later correspondence. There was no question of the existing scientists being commissioned as RAF officers: to put it bluntly, the service wanted those interesting jobs and saw them as berths which could be filled by run-of-the-mill officers 'e' for normal short tours on rotation.

The senior civil servants at Farnborough in May 1938 numbered six in the Superintendent and Deputy ranks: There were seven Principal Scientific Officers, fifteen Senior Scientific Officers (including Miss F.B. Bradfield, MA, AFRAeS) and fifty-two Scientific Officers (including Miss M. Lyon, MA, MSc, AFRAes). This makes seventy-four Scientists, not 132: But there were also seven Principal Technical Officers, twenty-four Senior Technical Officers, and eighty-four Technical Officers, 115 Technical Officer grades in all. This means 189 senior civilian posts which the Service engineers wanted for their career structure. The number did not rise significantly before the outbreak of the war. Most of the technical officers of all grades, and all the scientific officers, were graduates, the majority with higher degrees or other postgraduate qualifications. Bawdesey, not yet proclaimed as the Radar Research Establishment, had a Superintendent of Research, four Senior Scientific Officers, thirty-two Scientific Officers and Junior Scientific Officers, and one 'Executive Engineer' on a pay scale between £600–800 pa (Tables 21 and 22).

A critic of this proposal simply said there was no need for scientists at all. Engineer officers were mostly employed on administration in technical sections and on the maintenance units. A large number of maintenance engineers were needed, and a few for liaison with industry. Apart from that, the present system was ideal, with the engineering-trained pilot working with the promoted flight sergeant.

The last shot in this battle came from one who argued that the best way to ensure supervision would be to commission all Halton apprentices on graduation. This would remove the morale problem among apprentices, who saw chances of gaining Cranwell Cadetships or of making sergeant pilot as low, and the chances of normal promotion as not much better. If necessary, it should be possible to recruit a small number of university graduates as officers. They could serve under the ranker officers for a few years, to learn the ropes, and then should be sent to Farnborough: for life?

This was not such a long shot as it may appear. After all, the Halton entrant was educated to the same standard as the officer pilot entrant. An eye count over the Honours Board at Halton suggests that about one half of the pre-war apprentices were in the end commissioned.

Indeed, what emerges from these articles is the picture of junior officer pilots in a thorough muddle as to what they wanted engineer officers for, although all agreed they would be a good thing. The subsequent actions showed that Their Airships in Kingsway were equally confused. They were unable to distinguish three different kinds of work for which men with an engineering background and service training would be necessary.

First, there was the problem of supervision at the front line, the administration of the airmen on the flights, from fitters I to aircraft hands. The Air Staff tried to solve this first, by announcing that a number of senior non-commissioned officers in the technical trades would be offered commissions. These commissions would be on limited terms. Warrant officers and flight sergeants would be commissioned as flying officers, and would serve throughout their new engagements as flying officers; there would be no promotion. The *Estimates* speak of them as commissioned from among warrant officers. Such limited commissions have been common since then under various titles, such as the Branch Officer scheme of the 1950s and 60s, and derive from the old Army system of the Quartermaster and *Honorary* Lieutenant which saw H.H. Kirby into the officers' mess. They also resemble the early naval category of Gunners, Boatswains and Carpenters who were officers by warrant; if the RAF had been organized with a Gunroom Mess there is no doubt that they would have gone into it. And there is a strong flavour of the Imperial German Navy's Deck Officer system.

These officers' names appear on the *Air Force List*, after the chaplains and before the warrant officers, but not as a Technical or Engineering Branch. Their names appeared in separate columns under the titles 'officers commissioned for engineering duties', or for 'signals' or 'armament' or 'photographic' duties. The scheme was first rumoured in 1933. They appear in the *Estimates* for 1934, which allow for fourteen commissioned as engineers, four for signals duties and two for armament duties. By spring 1939 they had increased in the *List* to 125 engineering, forty-six signals, twenty-four armament and three photography specialists. And by now there were pay scales which envisaged their promotion as far as squadron leader.

There was also a need for service engineers and for scientists for research and development work, mostly at the Royal Aircraft Establishment at Farnborough. The tradition here was continuous from the days when Colonel Capper and Mr Cody had worked in harmony together to build and fly man-lifting kites and the first Army airship. The division between engineering on the one hand and physics and mathematics on the other is hazy in this environment. The Ministry tended to recruit Physical Scientists and Engineers into the Civil Service in different Classes, many as graduates but others in the lower grades with secondary education only. Of the posts which the service coveted, many were for lab technicians or bottle washers, for what were then called Junior Scientific Officers and later Scientific Assistants. In 1937 there were eleven GD officers at RAE at Farnborough, as well as twenty-five at Martlesham Heath, and twelve at the Marine Experimental Establishment at Felixstowe: however it is not possible to distinguish genuine R&D workers from the routine drivers airframe (Tables 24–6).

But the RAF had tried to build up its own body of better educated engineers in uniform. As well as the flow through Henlow, a smaller number of GD officers, that is of pilots, were sent to university, mostly to Cambridge, to read engineering. The peak of this enlightened work was reached in Frank Whittle, who entered as an apprentice, was awarded a Cranwell Cadetship to be turned into an officer pilot, and then sent to Cambridge to become an academic engineer. His achievement made the whole venture worth while. But genius or not, it took him an awful long time to reach Air rank.

In the 1934 *Air Estimates*, a scheme was announced in which University Graduates would be recruited direct from the University for permanent commissions, without going through Cranwell. They would be commissioned in the General Duties Branch, and taught to fly. They would then be employed on the normal duties of their rank, which put bluntly means that they would be pilots, although for part of their service they would be employed on, quite specifically, signals duties. The Air Staff were still dominated by the Trenchard doctrine that men who were recruited basically to be officers should be obliged to learn to fly. But down at the squadron level, the young officers saw themselves as pilots, they had joined to fly and any other task was an interruption. The macho image of the pilot and the doctrine of the 'Right Stuff' would be imparted early to the University entrants.

There was a third and different requirement for engineer trained officers separated from those for front line supervision and from the Research and Development field. Administration and planning would also require men who were at ease in engineering thinking, and understood the language of production. Somebody had to be in charge at station and group and command level, and above all somebody had to be in charge at the Ministry. If gifted men were to occupy these very senior positions, then it was no good leaving it to chance and hoping that men with engineering expertise keen on technical work would join to be pilots and somehow rise through the normal channels and provide a sufficient number at the senior ranks. The Navy had found out before the First World War that the only way to bring good engineers into the Fleet was to set up a separate Engineering Branch with guaranteed and protected lines of promotion to Flag rank.³ The RAF however persisted for too long with a system where flying was all, command of a flying unit was the goal for every officer, and the man who allowed himself to be sidetracked into engineering was taking himself out of the race for advancement. In the same way, the naval officer who chose flying was opting out of the race for promotion.⁴

Opposition to the formation of a separate Engineering Branch existed even among officers qualified in engineering. There were good reasons. Deep down in the service consciousness was the idea that the Air Force could only be commanded by pilots, just as in the Navy we find the belief that a fleet can only be commanded by a man who was once, early in his career, a ship handler. To allow progress to Senior, or even Air, rank to men who were not pilots would be to endanger the pilots' lives and allow those who did not understand air power to take control; there was a danger that if a separate engineering branch were introduced men might be brought into it who were not trained as pilots.

GD officers were pilots, who had joined the service to fly. They enjoyed flying. To transfer to another Branch meant no more flying. Also there was the suspicion that voluntary transfer out of active flying would leave a slur on their reputation. And there was money. A consequence of Trenchard's conviction that flying was a hard and dangerous task to which men would have to be bribed to resort was 'Flying Pay', that is an extra inducement of money paid to a man given a flying post. Although the regulation differed from time to time, there was also always the requirement that an officer claiming his flying pay must show that he had really flown the required number of hours in the year. This explains the retention for so long of 24 Squadron and of so many Station Flights: the qualifying air time need not be useful as long as it was done. After the war, the proportion of posts for General Duties Officers which were not flying posts increased greatly and income was affected by mere accident of posting: protests against this injustice led to flying pay being awarded to all officers qualified to fly whether they were required to fly or not.

When in 1940 the great step was taken of forming a separate Technical Branch, it included of course all the officers commissioned for engineering duties. Most but not all General Duties officers, that is pilots, who were qualified with degrees in engineering or the 'e' or similar symbol and who were employed in engineering posts were also transferred willy-nilly into the new Branch. Many of them resented this and regarded it as an injustice, or at best a piece of bad luck which deprived them of the hope of glory and distinction in the field which they had, as youngsters, chosen as their own. They could only reflect that, if you can't take a joke . . .

Terraine notes that in 1938 there was a suggestion that the fighter controller posts should be filled only by officers of 'the Signals Branch'. There was no Signals Branch or anything like it at this time, and this is not a mere quibble. The impression which Terraine has and which he tries to give to his readers is that there were then a large number of technical officers who were not pilots, and that it had been suggested that controlling fighters should be given over to these men.[4] What in reality we have is an attempt by the lobby of GD officers, that is pilots, with the 's' annotation, to ensure that all the squadron leader posts which were being created at the new Sector Headquarters should be reserved for them, thus improving their promotion prospects. One must remember that after the automatic hurdle to flight lieutenant was passed, no degree of merit could obtain an officer promotion, substantive or acting, if there were no empty post for him to be promoted into.

By the time the dust had mostly settled, in 1941, we find the Technical Branch *List* divided into separate *sub-Lists* headed Engineer Officers, Signals Officers and Armament Officers. In the same way the General Duties *List* had sub-divisions for Air Gunners, Air Observers, and Navigation Instructors. All were equal in the mess, but pilots were more equal than anybody.[5]

The Backing Groups

There had been other Branches than the General Duties since the very beginning. There had always been a Stores Branch (the Equipment Branch after 1936), since Honorary Lieutenant Kirby, RE, became Quartermaster of the Central Flying School. In 1930 there had been 303 Stores Branch officers, and oddly this had come down to 265 by 1934. It took to 1939 to work up again to 420. The *Estimates* in their *White Papers* never refer to changes in recruiting offers or terms of service, so we must assume that there was no difficulty in filling the officer ranks from commissioned rankers or from failed or unfit pilots.

It is interesting to note where these officers were employed. On the whole, there seems to have been one or sometimes two stores officers on every station, either at home or overseas: they are, confusingly, counted on the squadron strength at one-squadron stations. There were singletons in each aircraft carrier, and in a variety of schools where there was a flying task. But there were concentrations of Stores Branch officers, often reaching to double figures (this was still a very small Air Force) just where you would expect them, in the Stores Depots and in the Aircraft Depots, both in the United Kingdom and abroad.

There was a Stores and Accounts School in Kidbrooke, but it is not usually shown as having officers of its own, either Stores Branch or Accounts Branch: it presumably was essentially part of the resident unit, the Equipment Depot, and trained the airmen of this trade group. The main place where Stores Branch officers were trained seems to have been at the Home Aircraft Depot at Henlow. By 1937 there was an Equipment Training Unit at Cranwell, with one GD squadron leader on the staff and sixteen pilot officer students: Cranwell station headquarters (a separate formation from the RAF College headquarters) had six Equipment officers and six Accounts Branch officers on strength, and these presumably include the instructors on the course.

The Accounts Branch started its existence as the Stores Accounts Branch. These men were not Paymasters, but were basically purchasing officers working closely with the Stores Branch officers. There were 140 of them at home and abroad in 1930, 152 in 1934, and 201 in 1939. The distribution was like that of the Equipment officers, with one or two officers at every station, but with no concentration at the various Stores and Depots. Most of the officers seem to have been commissioned rankers, but the service wanted to attract professional accountants from civil life. Virtually every other *Estimates White Paper* announces a new scheme to increase the proportion of accountant officers with professional qualifications entering direct from civil life. The very slow increase in the branch may mean that none of these schemes worked, but on the other hand it is not real evidence.

The distribution makes it clear that the number of Accountant officers was not dependent on the numerical size of the Air Force in men, but on the number of stations and major units whose books had to be kept. The numerical increase here was just about right. The Accountant officers were not established to control or supervise men, as were the commissioned engineers, or to command and administer stores depots, in which things were physically handled, brought in, put on shelves, taken down again and shipped out, which was the task of about half the Stores Branch officers. They were recruited and commissioned to give them the status to carry out a specialized task, that is keeping the books of the service and of the individual stations to which they were posted. These were men commissioned to do their individual and very high-level hand work, just like the pilots and very unlike the officers of the army regiments.

They also resembled the officers of another long established group, the Medical Branch. These were not employed by the Crown to supervise nurses or medical orderlies (although they might do this incidentally) but to exercise their professional skill as doctors. As a result, the basic number of Medical Officers needed was related to the number of stations, since there was a need to have at least one doctor always on hand at every station, at least on call if not actually on duty. It was also, rather more directly than the number of accountants, based on the actual number of men to be treated. But the growth in the 1930s was on the same pattern as the growth in the number of accountants and equipment officers. There were 182 doctors in 1930, 185 in 1934, and 250 in 1939. There was also a tiny group of Medical Quartermasters, coming on the list after the doctors rather than after the Stores officers, rising from seven in 1930 to 9 in 1939. At first the RAF had relied on the Army Dental Corps for taking its teeth out, using borrowed Army officers borne on the permanent strength of stations, but on 1 July 1930 these were moved over to an RAF Dental Branch of twenty-six officers, who increased to twenty-seven in 1934 and reached forty-four in 1939.

Another professional group whose numbers were related more to the number of stations were the Chaplains. The Chaplain is not established to improve the fighting efficiency of a unit. He exists because, if the State sends its servants to duties in distant places, it is not seen as just to deprive them of the comforts of their religion. What Trenchard realized is that servicemen set great store by having their 'Own Chaplain' and their 'Own Doctor', even if they do not bother them overmuch. And they are not particularly thrilled with sharing these professional services with some other strange unit, and less than delighted if they find that to save money they are to be catered for on the side by the local civilian medical practitioner or religious incumbent as an appendage to his own practice or parish. The number of Chaplains was therefore also more related to the number of stations than to the number of men, and went from thirty-two in 1930 to thirty-five in 1934 and to sixty-four in 1939.

The hard fact which is not always appreciated by accountants and by time-and-motion-study experts turned into management consultants, is that a large number of trained professional workers are not employed for what they do all day hour by hour, but to be available when they are needed, because unless they are to hand in even a mild emergency the whole system will stop. The dentist may spend a long time with an empty chair, but when he is needed he has to be there.

There were a few other items on the list. Through the period there were always four or five officers in the Legal Branch. There was for long one Assistant Provost Marshal, an *Honorary* Flight Lieutenant. We also find almost always one Director of Music, and one Meteorologist. Even as late as 1941 we find only one Assistant Provost Marshal, but two Directors of Music. Somebody, in Sir Thomas Beecham's words, had to sing the bloody opera.

The Amateurs

During the last years of peace, two other branches appear on the *List* which were composed entirely of officers with Volunteer Reserve or Auxiliary commissions. It must be remembered that once conscription had come in, everyone entering the Service was enlisted or commissioned not into the RAF but into the Volunteer Reserve and until midway through the war wore the tiny brass letters VR on his lapels. Officers with Auxiliary commissions, like members of the Womens' Auxiliary Air

Force, wore a letter A. We may compare the American system, where men were enlisted or officers commissioned for the duration of hostilities not into the US Army, but into the Army of the United States. Since the Americans lacked the device of acting ranks, regular officers were promoted to higher ranks, and often to General rank, in the Army of the United States, while remaining in their original low positions on the list of the US Army. Such transfers were not possible under the British system.

The first of these branches was the Balloon Branch. The officers were not intended to go up in the balloons as in 1918, but to supervise and administer the barrage balloon units. The barrage balloons were kite balloons, or *drachen*. All officers of this branch held Auxiliary commissions, and were called to active duty just before the balloon actually went up. By 1944, the old Balloon Branch had been ingested into the Technical Branch.

The Administrative and Special Duties Branch of the Volunteer Reserve appeared in 1939. Originally this Branch administered nothing but itself. The early advertisements make it clear that it was to consist of older steadier men, preferably with flying experience, especially in the RFC, to carry out the process now known as debriefing, the questioning of crews after landing as to what they had actually done and seen on the sortie. The Special Duties were soon expanded to cover the men – few, for this work was mainly done by women – who ran the radar chain, or the Control and Reporting Organization as it was officially called.

By 1941, this Branch was sub-divided into a variety of subheadings. There was a sub-list of officers appointed for Administrative Duties, and others appointed for Intelligence, for Marine Craft, for Photographic Duties, for Assistant Provost Marshal Duties (presumably to help the one solitary regular APM), for Physical Training Duties, for Special Duties, and an intriguing group headed 'Unclassified'. The Physical Training officers were regulars: all the others were either Volunteer Reserve or, a very few, Auxiliary officers. But this organization was still settling out of an inchoate mass when, in 1939, first conscription arrived and then the RAF went to war.

Selection
Selection of these ground branch officers was less, than

systematic, given the applicant population. On the one hand, men were selected for commissioning because they applied for branches where they would be required to display some specific skill, like accountancy. On the other hand, men would be approved for commissioning by a small interview board on half an hour's acquaintance, and largely it would appear on the basis of social and, to a lesser degree, political, acceptability.[6] This had not advanced far from the selection of infantry subalterns in the eighteenth century, except that these men were not required to put any money down.

The Army had a different system, eventually. Originally in the early days of conscription, units were required to submit the names of men recommended for commissions, and they too were approved as a result of an examination of the documents and a short interview. The results were not encouraging, as any Commanding Officer with an eye on his own prospects would recommend young gentlemen of amiable aspect with their Cert 'A' and of no particular ability, and would certainly not send up for approval the tough and efficient Corporals on whom the running of his unit depended.

After the French campaign of 1940 the Army were much impressed by the superior performance of the Germans at unit level, and wondered how they got their young officers. The General Officer Commanding Scottish Command, Sir Andrew Thorne, had been Military Attaché in Berlin in 1936, and remembered the report of two RAF officers who had been attached to observe the German manoeuvres of that year: a not uncommon occurrence in the military intercourse of countries which were not at war with each other. (The famous visit of General Milch to the RAF must be considered as simply part of this conventional sequence.) They described their experiences in the *RAF Quarterly*.[7]

One evening, during the manoeuvres, they dined with a chance acquaintance, a German official. He told them, in strict confidence, of a new system the German Army had devised to select officers. There were, he said, fifteen Service Psychological Examination Centres, each with eighty-four psychologists and fifteen senior officers. Here candidates were brought and observed in teams of eight, going through all the routines of leaderless and leadership tasks, problem solving, and shallow psychological probing with which British candidates for commissions became familiar later. The observers were not physically present at these exercises, but made their decisions entirely on films and recordings. 'In 95 per cent of all cases controlled,' the RAF officers were told, 'reality has coincided with theoretical judgement.' How this validity was measured was not explained, but it looked impressive.

There is no other evidence that this procedure was actually followed. While the War Office Selection Board procedures, as they became in British hands, entered into folklore, there is no record, as one might expect, in German memories or literature, let alone in history, of the German Army selecting officers in any other way than by the traditional method used before 1870 by the Prussians. In this the promising junior NCO (including usually the young Junker who had joined up in order to become an officer in the tradition of his class), once he had gained his stripe, would be sent to the Divisional School for officer training. He would then be returned to his unit, where his suitability for commissioning in that unit would be considered by the other officers: he might be blackballed. After 1934, Party membership was an advantage, but was not essential: competence was the dominating consideration.[8]

Nevertheless, General Thorne remembered the system as he had been told about it, and, at the beginning of 1941, began to copy it in Scottish Command. Psychologists were scarce, but he found two, and made up the rest of the establishment of serving officers with war experience. The system proved effective, and very soon became the standard method of selecting British Army officers.[9] It worked mainly because the Army no longer looked for recommendations, but simply put everyone with School Certificate through. Under these conditions, with a very high proportion of candidates to places available, the suitable men leaped at once to view. And the troops believed in it because everyone had a chance.

The RAF did not introduce this system till after the war, first to select from the mass of wartime officers who applied for permanent commissions, then for applicants for Cranwell Cadetships, and after 1953 for all officer applicants. The system was taken over by the Civil Service Commission, by the Church of England, by the American CIA under its earlier names, and by a variety of civilian enterprises under the influence of management specialists. It has been difficult to validate the procedure, although there is always a gut feeling that even if it is open to criticism at least it must be better than a single half hour interview. Group Selection Procedures in general have become almost a sub-industry. Yet it seems possible that they all derive from a rather heavy evening in a German bar where an anonymous German psychologist went on to two innocent RAF officers on the lines of 'this is how it ought to be done, and in particular wouldn't it be nice if the film industry were advanced enough, in 1936, to make sound films of candidates in session without disturbing them with lights and cameras and microphone booms and process them overnight.' As so often in the history of the RAF very doubtful motives and origins led to fruitful results.

Women
Although Trenchard was no longer there to guide them, the RAF now made a massive social innovation. Sykes had been asked to make it possible for the Army to use women in uniform to replace men. He had invented the Womens' Auxiliary Army Corps. It was fairly obvious by 1938 that when war started the Army would again do something of the same kind, and use women in uniform to do what one may call the domestic jobs where they could replace men, the jobs of cooking and cleaning and filing and typing and driving. The RAF went a step further, using the Women's Auxiliary Air Force.

The posts in the radar stations and sector control rooms were given over to women. These were not domestic posts, nor any kind of support structure for the real fighters and workers. These women *were* the real fighters and workers; they were an integral part of the air defence of the country on a level with the aircrew. They were even more directly concerned in the air war than were the servicing crews on the stations turning round the fighters and who could just as easily have been posted to turn round bombers or trainers instead. The RAF brought women into war on equal terms with men in a way which had never before been dreamed of. They were as much direct combatants as were the Marines who did the same job in the plotting rooms of battleships. War had changed, and the Air Staff saw the changes coming.

Notes and References

1. AVM Sir Philip Game, 'Service in the RAF', *RAF Quarterly*, January 1930.
2. The correspondence in the *RAF Quarterly* began in October 1936 and continued till January 1939.
3. Michael Lewis, *England's Sea Officers*, George Allen and Unwin, 1939, *passim*.
4. John Terraine, *Right of the Line*, Hodder, 1985. Terraine does not adequately understand the organizational revolution of early 1940 which transformed the RAF from an Army into a Navy.
5. See T.T. Patterson, *Morale in War and Work*, M. Parrish, 1955, esp. pp. 31 seq.
6. Hugh Thomas, *John Strachey*, Eyre Methuen, 1973.
7. *RAF Quarterly*, 1936, p. 508.
8. M.M. Mason, *Rise of the Luftwaffe*, Cassell, 1975, p. 260.
9. Vernon and Parry, *Personnel Selection in the British Forces*, University of London Press, 1949, pp. 54, seq.

Chapter 12

Plans and Pilots

Questions
All we have said so far leads up to asking some very specific questions. Some are basically political, being concerned with the plans which the government had promised at various times as targets for the expansion of the RAF. These plans are always stated in terms of squadrons. The first question addressed to the Air Staff by the taxpayer through the Government of the day is always: have you fulfilled the plan for which we voted money? Are the squadrons there?

But squadrons may be merely number-plates. Squadron numbers are only real when they are filled out with pilots and with aeroplanes, positioned on airfields. The most squadron numbers do is to suggest the tactical organization which will operate those aeroplanes with those men. The question to be asked is, realistically, as before: how many sorties could you put up immediately, Air Marshal, if war broke out at eleven o'clock tomorrow morning? And meanwhile, what happens to the Empire? The answer was never simple, but depended on a variety of considerations.

And there is a third question, to be asked of the Air Staff by the politicians: or perhaps a set of questions, which again invite a number of answers. If war breaks out tomorrow, are you going to be able to do anything with the aeroplanes and pilots you have raised to make sure that the enemy's bombers do not get through, and that ours do? Because if, Air Staff, you have guessed wrong and cannot stop the bombers and get ours through, then we are in a bad way.

Plans

The lettered Plans we have heard about (Table 12). It is less than easy to interpret these plans for expansion. They are all for the United Kingdom based Air Force. It is never clear whether they mean only regular squadrons, or whether they count in also the cadre and Auxiliary squadrons: this may make little difference to the final order of battle well on in a war, but it does affect the effort it is hoped to put up on the first day. In none is there any provision for communication squadrons, yet 24C continued to exist and to operate throughout the period.

When Plan 'A' was drafted, in the summer of 1933, the stations in the United Kingdom housed thirteen Regular fighter squadrons, thirteen bomber, one torpedo bomber, five Army cooperation, and four flying boat, or thirty-six operational squadrons. There were also 24Comm and the two purely theoretical bomber squadrons at Martlesham Heath. There were five cadre heavy bomber squadrons and nine Auxiliary light bomber squadrons. Thus the eye of tolerance and hope could claim that there existed fifty-three squadrons in the home-based Air Force (Table 14).

The first step towards fulfilling Plan 'A' was the obvious one, practical but without obvious relevance to the fulfilment of the paper plan. Quietly and without publicity for what was really happening, the number-plates of 15 and 22 squadrons were taken away from Martlesham Heath, and given to two brand new squadrons apparently formed fresh, a light bomber squadron at Abingdon, and a torpedo bomber squadron at Donnibristle. But almost at once the existing torpedo bomber unit, 100TB, at Donnibristle, vanished and reappeared at Singapore. You can do anything with figures, or at least with number-plates. There was thus a net increase, unannounced, of one operational squadron in the UK.

In 1934, an exercise was carried out in which 4 Army Cooperation squadron were transported from their base at Farnborough to the Packing Depot at Sealand and prepared and packed with their aeroplanes (Audax), equipment and men for going overseas. Without going anywhere they were then sent back to Farnborough. It is not known how the men were crated, or what they said when they were told to go back home again. The exercise, however, had its uses, because the following year, when the Abyssinia crisis developed, four squadrons were sent to the Middle East, 3F and 29F to Egypt, and 12B and 41F to Aden. To make this move possible, they had to fill out their ranks with the aeroplanes of 17F and men and aeroplanes from 23F, so that the home defence force was, at least at first, effectively depleted by six squadrons. These squadrons were restored, or dribbled home as units, one by one, during 1936.

There were some signs of real expansion late in 1934. 142B was formed at Andover, and 65F at Hornchurch. Three Auxiliary light bomber squadrons, 600, 601 and 604, all at Hendon, were each re-designed as fighter squadrons, although they retained their old aeroplanes, Harts for the first two and Wapitis for 604.

During 1935, three new fighter squadrons and, late in the year, six bomber squadrons appeared. In 1936, six bomber squadrons were formed in the first quarter of the year, and then in the second half of the year, another five. In the same year, four fighter squadrons were formed. During the first six months of 1937, twenty regular bomber squadrons were formed, and seven fighter squadrons.

No further new bomber squadrons were formed in 1937, and only three in 1938, to bring the strength up to the intended sixty-eight. And no more new bomber squadrons were formed till the end of 1940. One new fighter squadron was formed in the autumn of 1937, and one in 1938. But during late 1937, the bomber force lost one Auxiliary squadron converted to a fighter title, three in autumn of 1938 and five, with two of the old cadre squadrons, in the spring of 1939. The next fifteen fighter squadrons to come arrived in the autumn of 1939, after war had started.

Thus, in the spring of 1937, when Plan 'C' was due for fulfilment, the RAF had, in the United Kingdom including Auxiliary squadrons (the cadre squadrons had all now been re-titled Auxiliaries) a total of fifty-eight bomber squadrons, and thirty fighter squadrons. There were also five Army co-operation squadrons, six flying boat squadrons, six general reconnaissance squadrons, two torpedo bomber squadrons, and of course the one communication squadron. This does not amount to fulfilling Plan 'C'.

A year later, in spring 1938, the total had risen to sixty-seven squadrons of bomber, thirty of fighters, ten for Army co-operation, six of flying boats, two of torpedo bombers and, still, the one communication squadron. Still Plan 'C' had not been fulfilled, at least as far as fighters were concerned. Perhaps if the long-suffering 24Comm squadron, not otherwise accounted for in any of these plans, was rated as a bomber squadron, it might be claimed that Plan 'C' had nearly had some success. During late 1938 it is true one new regular fighter squadron was formed, but it was to replace a squadron sent to Egypt and therefore meant no increase in UK based strength.

The lesson stands clear for the general historian. At the time of the Anschluss in spring 1938, the Service had reached a peak strength in units, that is an organization, which hardly increased through the following eighteen months. In terms of squadrons, the RAF was as large now and at the time of the first Czech crisis (the Munich Crisis) as it was at the second Czech crisis and at the outbreak of war. As to whether it was ready to fight as early as September 1938, we must go a little further.

We must remember the nominal squadrons at Martlesham Heath. Were all these new squadrons real? In fact they were not.

Bomber Command had looked at the question of replacements for battle losses, and at the problem of training men when they came from the flying training schools, where they might possibly have reached the sophistication of a Hind, to wrestle, literally, with the controls of a Whitley and command of a crew. Films and comfortable holiday flying have seduced us into thinking of handling wartime combat aircraft as on a par with sitting in a saloon car talking on the Cellnet. Flying a military aeroplane at the time was perhaps a little easier than it had been in 1914, but it still resembled at best driving a badly suspended heavy lorry fast through traffic on bad roads in winter, while conversing with the traffic police through megaphones. Weight was not wasted on soundproofing. We are reminded of the Japanese who did not issue mosquito nets to their aircrew in the Pacific, since real men would not be worried by the trifling bites of insects, unlike the effete American pilots who were given nets and so escaped the malaria which played havoc with the Japanese air effort.[1] We can never know how many aircrew were lost because at the end of a long sortie the men were simply worn out with noise and vibration and cold and smell and stuffiness and too exhausted to find the airfield or to land on it. Working under these conditions was a skill to be learnt, and taught.

The schools to bring pilots from the Hind to the Hampden or Whitley were set up, but were not called schools. The politicians wanted squadrons, in numbers, so numbered squadrons they should have. The first impression one has during this period of rapid expansion in terms of squadrons, is that as soon as a new station was at all usable, then a station headquarters and two squadrons were created and put into it. Things were not quite as simple as that, but it is largely true in terms of number-plates. Of the new bomber squadrons, twelve were designated as 'Group Pool' squadrons. They were not operational squadrons, and in April 1940 they were all formally disbanded in the sense that their number-plates were taken away. The stations were re-titled Operational Training Units, each internally administered as two squadrons as Martlesham Heath had been. Most of these squadrons were re-formed during the next two years, in the sense that the number-plates were given to new bomber squadrons.

This manoeuvre allowed a larger number of squadrons to appear on the *List* and on the information handouts. However, the actual number of operational bomber squadrons available at the beginning of the war, Regular and Auxiliary, was not sixty-seven but fifty-five.

Not all even of these squadrons were available for a conventional bombing campaign. Two separate forces were organized to accompany the Army to France. One was the Air Component of the British Expeditionary Force, and was made up of four Army co-operation squadrons (Lysanders), four bomber/reconnaissance squadrons (Blenheims) and four fighter squadrons (Hurricanes); this was under the orders of the C-in-C BEF. The role of Army co-operation units was firmly laid down: they were for reconnaissance and artillery spotting, and not for ground strafing. Ground attack was the task of bombers when they could be spared from their proper duties.[2] This force closely corresponds in its task to the Military Wing when it went to France in 1914.

The other was the Advanced Air Striking Force of ten bomber squadrons (Battles), under the command of C-in-C Bomber

Command. This brings down the number of Regular bomber squadrons available in the United Kingdom, to forty-one, and the number of fighter squadrons available in the United Kingdom, including the Auxiliaries, comes down to twenty-five.

In fact, the evidence is that what the Air Staff achieved was what they intended in the first place: they completed Plan 'A'. The figures make us go farther than that. We have to think of the Air Staff in 1933 working out a war plan, which would make use of just over forty home-based bomber squadrons, and about twenty-five fighter squadrons, and sticking to it, adjusting their titles to satisfy the political demands. In addition, presumably with careful attention to what the generals wanted, they provided over and above the original plan a further force of about twenty squadrons to accompany the Army on its foreign adventures.

Any historian who writes as if he believes that the Air Staff tried to fulfil any of the Plans other than their own simply has not looked at the easily available primary evidence. We should examine the reliability of any of his other statements.

In November 1939, two Auxiliary fighter squadrons, armed with Gladiators, were sent to France, but must be related to the fifteen new fighter squadrons formed at the same time. On 15 January 1940 both AASF and the Air Component were brought under a new command called British Air Forces in France.

So when the politicians asked, if they ever found out that this was the right question, and used sufficient force to ensure a reply, the question, 'Have you provided the squadrons we have voted you the money to provide?' the Air Staff would have had to answer, 'No. Instead we have provided what we offered you in the first place, an Air Arm organized to carry out a coherent war plan, and big enough to make some difference in a war.' The German Air Attaché would not necessarily have known about the Group Pool idea, but in his reports to his superiors British air strength might have been somewhat exaggerated.

Whatever politicians and historians may believe, a war is not successfully fought simply by massing together the biggest army possible. Space as well as time is of the essence. Too big an Army in too small a space simply gets in its own way. Too small an Air

Force would be swamped. But too large an Air Force would simply mean desperate overcrowding for airfields at the crucial times of dawn when the night bombing details would all be trying to get down together, and mid-morning when the fighters sent up to counter the dawn attacks would be trying to get down again, all at once with no airfield radar and no real Air Traffic Control. 11 Group was the right size to operate from the airfields in its area: a bigger force would have made the air space available for the battle seem like Gatwick on a Bank Holiday. Plan 'A' provided moderation, not usually a virtue of politicians.

Crews

This brings us to the problems of getting enough pilots. There are several factors to be considered. First, training pilots is an affair of geometric progression. One instructor can teach at most four pupils, on the basis of an hour's sortie each per day, with about half an hour for instruction before the trip, and about the same time for debriefing after it. That is what the task charts would be based on. In Great Britain, with the normal conditions of day length and weather, it is not always easy to get in that much instructing every day: or any day.

The instructor is not there simply to be a safety guy and sit immobile while the student lands the aeroplane and lands it wrong – again. The instructor has to be taught to teach, and although sometimes CFS drifts into being only a place where satisfactory pilots are taught to fly better, they have also to be taught the patters and all the rest of the Smith Barry gospel. If a decision to expand were taken in November 1933, it would still be a considerable time before enough new instructors could be trained to teach the new pilots.

There was another time constraint. Most of the new pilots would be recruited on short service terms of six years active. Therefore there was no point in recruiting them more than six years before the war was expected to begin, and preferably the bulk would have to be brought in about two years before, which means during late 1937 and 1938. The bulk, that is, of the men needed to fill all the aeroplane seats at the beginning of the war. Recruiting and training would have to be continued from then up to the outbreak of war and well after it, since the one thing the Air Staff knew about air war was that it was inordinately expensive in men. And in educated men, men with School Certificates, a drain on the pool which was also going to have to supply officers for all three services, and all the skilled men for the services and for industry. Providence, having invented the tools of the Officers' Training Corps and the Somme Offensive to destroy the cream of one generation, now invented a tool to do the same job for the next generation: it was called the Bomber Offensive.

The first move to increase the number of pilots available in emergency was tentative and cautious. In 1934 it was announced that forty Direct Entrants a year to the Volunteer Reserve would be trained as pilots, in the rank of sergeant. Given the difficulty of finding men who could spare the time from their jobs to carry out even a part of their training at service flying training schools, and assuming the normal wastage of 50 per cent in training and a little more wastage from lack of continuity and friction between the reserve and civil life, perhaps the Air Staff were counting on getting fifteen sergeant pilots a year to fly in the cadre squadrons.

In January 1935, came an announcement that the two University Air Squadrons at Oxford and Cambridge were to be increased in size above their establishment of seventy-five students each. All candidates were to be accepted. 'A part' of their training was to be carried out at Cranwell. These two Air Squadrons were formed originally not as recruiting agencies, but to give flying training to young men who were seen as about to take their places in the upper ranks of the Establishment, especially as civil servants (Administrative Class only, of course), public school masters, lawyers, and on the board of large, usually family, firms: eventually many of the latter two groups would go into politics. These men were seen as providing the RAF in half a generation's time with the kind of political lobby already and traditionally available to the Army and Navy. If a few of them wanted to join the Service, that was a kind of bonus.

But this new development abandoned the original academic and social exclusivity which had been a feature of the squadrons, and jettisoned the long-term aim in favour of immediate results. Expansion of the instructing and support staff would take some months. The freshmen of October 1935 would for the most part be taking their degrees in June 1938, and could be recruited in numbers for the campaign which was anticipated in or after spring 1939.

In late 1936 another scheme was announced. A number of Volunteer Reserve pilots were to be recruited, and commissioned. They would be drawn, it was hoped, from those men who wanted to spend a productive year before going to University or into the professions. The whole year was needed for training. The first half would be spent at a civilian flying training school, and the second half at a service flying training school. At the end of this time they would be awarded their brevet and pass to the Volunteer Reserve in the rank of acting pilot officer on probation. In other words, as we shall see, they would be trained exactly like the regular short service pilots.

For these two additions would only be drops in the ocean. The great mass of the pilots needed for the home Air Force of eighty-four squadrons (or 123 or 145 or 168 squadrons depending on which plan you believed in), would have to be brought in by the current system of Direct Entrant (Short Service) commissions. As we have seen, this system called for recruiting men with their military training completed at public school, bringing them to Uxbridge for a fortnight's indoctrination and documentation and to get their uniforms, and then sending them to a flying training school for about a year, to receive at least eighty hours flying instruction of which twenty were to be on 'service', that is operational aircraft. This last requirement dated back to the regulations of 1916, when it was assumed that a man could be sent direct from flying training school with his wings, to a squadron. In the 1930s the 'service' aircraft with which the schools were equipped was usually the Hart or later the Hind. At this period, about half the squadrons were equipped with the Horsley and its direct descendant the Hart or its developments, the Demon, Audax, Hardy, Hector, Hind, and so on. Therefore the requirement was realistic, but not later, after the monoplanes appeared.

The service flying training schools were not going to be able to cope with the flow of trainees expected. In 1936 the system was changed, and the changeover took at least a year to work itself through. From now on, the pilot applicant would be sent to a civil flying school for eight to ten weeks, as a civilian and without uniform: by 1939 thirty-four civil flying training schools appear on the *List*. This, as we have seen, was the scheme which the Volunteer Reserve officers would join.[3] We may assume that this stage would suffer about 25 per cent wastage. The survivors would be sent on to Uxbridge for an introduction to the RAF, and would only then, we must assume, be attested and commissioned. Immediately on commissioning the student would be sent to a service flying training school, for an intermediate stage of thirteen weeks, ending with the award of the Flying Badge. This stage would include night flying instruction to solo, and instrument flying under the hood. Only CFS possessed that wonder, the Link Trainer, and were trying to get another one. Now the young officer would be allowed a fortnight's leave, before an Advanced Stage of thirteen weeks, also at a flying training school.

The number of service flying training schools was also increased. Up to the expansion, there had been always 4 Flying Training School in Egypt. This was retained. When the war started, this school, or at least its number-plate, 4FTS, was moved down into Southern Rhodesia, as part of the Empire Air Training Scheme. When the Rhodesian Group was shut down in 1954, the number-plate was brought home and attached to the advanced flying training school at Valley. In the United Kingdom, there had up to 1935 been three service flying training schools. 1FTS had always trained all the naval pilots, although some RAF and Army pilots were also trained there from time to time.

The number of flying training schools in the UK remained at three, with occasional changes of location, until 1935, when two more were opened. By spring 1937 there were ten, and this number remained stable. The numbers and ranks of the students are, however, worthy of comment. In the 1920s, when the young gentlemen were commissioned at Uxbridge before entering training, the trainees were divided into NCOs and pilot officers. After 1936 the flow of pilot officers begins to taper off. A new rank appears as that of the normal entrant who made up the bulk of the pupils, that of Acting Pilot Officer. These are presumably the Volunteer Reserve officer candidates, or an unannounced change in policy may have meant that the regular applicants were now commissioned at Uxbridge as Acting Pilot

Officers (leaving the question of what was their substantive rank for endless bar discussion in later years).

By spring 1938 there were at any one date about 500 RAF officer students at the service flying training schools for a course of roughly half a year, which would mean that the year's intake would be of the order of 1,000. In the absence of detailed information, we may safely assume that the wastage would be about one third, and therefore that the service would be taking in to the squadrons about 700 officer pilots a year.

These estimates of the size of the flow into the squadrons are of necessity inexact. There was an interval of at least a year while the original entrants beginning their flying at the FTSs were completing and the first entrants were coming from the civil schools. They are more complicated by the fact that as far as the schools are concerned, while the number of NCO pilots on the instructional staff are always given in the *List*, the non-commissioned pupils are ignored. However, they are the calculations we must assume the German Air Attaché made.

After August 1938 the *Lists* cease to report the number of NCO pilots in the squadrons or on the staffs of the flying training schools. It may, however, be a justifiable assumption that the old ratio of one NCO to every three officer pilots continued.

The Central Flying School *List* presents a different problem of interpretation. In theory CFS existed to train flying instructors. The *List* always gives the names of the officers and the number of NCO pilots on the directing staff and the names of the officers under instruction. For all the period between the two wars the number of directing staff remained stable at about two dozen, and the number of officer students was never more than fifteen, often down to four. It would appear that expansion never happened here.

We must assume that regular officers and NCOs who were sent to CFS for training as instructors were merely attached from their home units, probably from the flying training schools to which they were posted; these temporary attachments would not be recorded in the *List*. The tiny numbers of officers under instruction at CFS – usually around five – were presumably being trained not for instructing at the FTSs but for instructing at CFS itself, after formal posting in.

The mystery remains of how CFS managed its increased task.

The standard course for instructors lasted eleven weeks, and was made up of forty pupils, of whom about a quarter were NCOs. The minimum pre-appointment requirement for instructors at this time was a total of 1,000 hours flying time. In August 1938, the course was reduced to nine weeks, and its size increased to fifty students. And the great decision was taken.

CFS would train what are usually now known as 'creamed-off' instructors, that is men taken straight from the flying training schools and taught to teach flying without any squadron experience. And since the first stage of flying training was to be carried out by civilian flying clubs and commercial schools, their instructors too had to be trained or at least refreshed, which was done on a very short course at CFS.

CFS was now responsible for writing Pilots' Notes – that is the instruction book – for each new aeroplane as it came into service. There was an Examining Flight, nowadays a Wing, which played the part taken in the normal education system by Her Majesty's Inspectors of Schools. This Flight toured both the flying training schools and the operational units to check on flying and instructional standards. And somehow while the RAF was expanded, this was all done without increasing the established or actual officer strength of the School. The standard history tells us how busy everyone was at CFS at this period, but does not account for how all this work was done by so few.[4]

After Flying Training

Most pilots were posted direct from flying training to a squadron. A favoured forty-eight every year were posted to flying boats. Between flying training and squadron service they would spend twenty-two weeks at Calshot, and do 145 hours flying, including a 'cruise' of fifty-six flying hours away from base.[5] The first act of a trainee arriving at the School at Calshot was to hang his brand-new hat on some convenient projection on his boat where it lay in the river, till the wind and the spray turned it a good distinctive green colour; the boat people, 'Greenies', considered themselves a *corps d'élite* and demanded a distinction in dress much more elegant than leaving a tunic button undone. Another small group was sent to the School of Army co-operation to learn the task of Army flying, especially artillery spotting, using either the Hector or, by late 1938, the

Lysander. About half of these pilots were Army officers in khaki, and considered this distinction enough. The remainder of the trainees went direct, with something under 100 hours flying, to a bomber or fighter or general reconnaissance squadron. There were at the later stage the Group Pool squadrons, but no specialized training schools for fighter pilots.

We may conveniently divide the new squadrons produced during 1937 and 1938 into fighter and bomber. Later in the prewar period a number of the day bomber squadrons were redesignated as general reconnaissance squadrons, already armed with Ansons by 1937. The remaining 'genuine' bomber squadrons may then be divided into the day or light bomber force – Hinds, Blenheims, Battles, according to date or place – and the night or heavy bomber force – Virginias, Heyfords, Hampdens, Whitleys. Or in more modern parlance, into a Tactical Bomber Force and a Strategic Bomber Force.

Navigators

As far as the strategic bombers and the general reconnaissance force was concerned, it was in late 1937 that the RAF realized that it didn't know where it was going. Traditionally navigation was done by map reading and by the judicious use of *Bradshaw's* railway timetable. The heavy bombers had positions for two pilots, one of whom did the navigating. The gun positions were worked by airmen who had qualified as air gunners and wore the flying bullet on their sleeves, were paid extra for the qualification, worked normally at their trades, but who were to hold themselves ready to fly whenever needed. They got flying pay for the time spent in the air but flying was a secondary duty.

We get a few clues as to how this went on. In 1932, *Flight* records the inquest on the crew of a Virginia which crashed at Upper Heyford, after being impeded in the circuit. The captain was a sergeant, the second pilot a flying officer, the two gunners were airmen, and all were killed. There was also a corporal on board as a passenger, and he was unhurt, thus proving an exception to the usual rule of the innocent bystander.[6]

In 1939, Basil Embry took over a Blenheim squadron at Oakington. Among the available qualified aircrew he chose Corporal Whiting as his navigator, and Leading Aircraftman Lang as his gunner, for no other reason, he says, than that he liked the look of them. After some time and trouble he was able to get the corporal made up to acting sergeant, unpaid, because he thought it was wrong for security reasons that the station commander's navigator, who would know about operations the night before, should have to sleep in the barrack block among the other airmen.[7] This corporal would have been trained as a full-time observer in a scheme started in 1934: his duties would have included map reading and course plotting as well as bomb aiming, tapping the key and making the tea as the poet has it. The air gunner would have been a part-timer, although how long this system lasted in wartime is debatable.

There had been for many years a school of navigation, the School of Air Pilotage, giving a one-year course for pilots. Those who succeeded were given the annotation 'n' in the *List*. Most of the very few 'n' pilots were employed on staffs, and at the big coastal bases like Gosport and Calshot, although there were always a few in squadron service in flying boats, or in Mesopotamia where *Bradshaw*'s writ did not run. In 1937, the annotation 'sn' appeared to denote completion of a short navigation course, and by the end of 1938 there were enough pilots with this annotation available to allot one, not to each squadron, but at least to each heavy bomber station, presumably as what we would later call Station Navigation leaders, and one to each flying training school. As it was not practicable to train every pilot to this standard, and in any case the navigating job took a man out of the main stream of piloting, the service took a drastic step at the beginning of 1938, and created a new Officer Aircrew category. This was done quite practically to satisfy and reduce the pilot requirement by training a cheaper substitute for one of the pilots previously needed in a big aeroplane. Men were to be recruited on the same short service terms as pilots to become Observers.

The Observer category had existed, especially in the RNAS, and also in the Independent Air Force as late as 1920.[8] It seems to have died out largely as a result of Trenchard's insistence that every officer would have to learn to fly, and the rank-title of Observer Officer, as an equivalent to Flying Officer, had also disappeared. The legend in the Independent Air Force was that every observer was a picked man, since he had been picked as incapable of becoming a pilot. It is true that many of the early

Observers were failed pilots, but not all, or even many, of them. The School of Air Navigation was formed in early 1938, and the first entrants were a dribble of eleven junior officers, and one squadron leader: these may well have been retread pilots, for medical or other honourable reasons. The first course of Direct Entrants numbered sixty acting pilot officers, that is men just entered and commissioned a week or so before the course began, not failed pilots: it started on 13 June 1938, and lasted for seven months, when the next course of acting pilot officers arrived to replace it. From now on it was not possible to say that every GD officer was a pilot.

The Manning Crisis
From the middle of 1938, a state of confusion developed in the squadrons as we have noted at Upwood. The manning table shows why. The Service was beginning to acquire a large number of young pilots. The old pre-expansion fighter or day bomber squadron of 1934 had about two dozen officers and sergeant pilots, slight variations being unavoidable in a living organism as the service was. The squadron was commanded by a squadron leader, with flight lieutenants as flight commanders and as an engineer; all these were men with from six to ten years service. There would be, usually, two or three pilot officers, accounting across the service for a year's output from the schools, and still regarded as sprogs. The rest, the journeymen pilots who did the work, would be eighteen or twenty flying officers and sergeant pilots, with anything from two to six years service in the flying game.

By the middle of 1938, the *Lists* show that the typical day bomber or fighter squadron had a squadron leader, one flight lieutenant or one flying officer but not both, and a score of pilot officers and sergeant pilots, all very new out of the flying training schools. A heavy bomber squadron also had its wing commander on top of that, but almost never the two squadron leader flight commanders. There was nobody in charge. Several authors describe the scene. According to Embry the squadron commander was accepted as being much too busy in the office to go into the air.[9] The same applied to his deputy, who might be a flight lieutenant or a flying officer. Command of the squadron in the air, that is Squadron Air Operations, would devolve on the most senior pilot officer, who would have less than two years experience in the service. The flying programme would be unplanned, and men would fly at random, when aircraft were available.

Sykes' nightmare had come true, and it was partly Trenchard's fault. The Air Force had run out of officers, that is officers in the working rather than in the social sense. But the shortage of experience at this date was implicit in the original manning plan, and in the expansion programme from the beginning. Somehow, while Trenchard had been planning and building a Service to last for ever, he and his successors had lost sight of the vital four minutes.

It had been seen coming. In 1936, junior officers were worrying about the shortage of flight lieutenants. Something was done about that.[10] Up to then, automatic time promotion brought advance for the flying officer qualified by examination after four years in the rank. In that year, the qualifying time was brought down in two successive steps to first three and then two years service in the rank. But the new flight lieutenant would have only a small rise in pay to go with his new rank title, and would not be paid as a flight lieutenant on the old scale till four years after he had been first promoted to flying officer.

There was here a jumbled recognition that merely putting an extra ring on a man's sleeve was not the whole answer. What was wanted was not more flight lieutenants but more flight commanders: the change in promotion time still left a shortage of hard and experienced men used to the air and to military life and therefore fit to command formations in garrison and in action. Probably on most squadrons by the middle of 1938 the men who were best qualified to be flight commanders were flight-sergeant pilots with three or four years in flying, but of course they could not be used.

The zone for promotion to squadron leader seems also to have been widened at the bottom end. The tough hardened flight lieutenants were promoted in large numbers to be squadron leaders; none were left to be flight commanders. These men with air experience and with knowledge of the business of a squadron were needed for a mass of other posts which were necessary just to run a much larger Air Force than anyone had before thought possible.

The Empire

The overseas strength of the Service in terms of squadrons had remained stable through most of the interwar period; or at least the slow increases in Egypt and the Far East were in proportion to the general growth of the Service.

However in the later part of 1938 and early 1939 a good deal of rearrangement and reinforcement went on. Two bomber squadrons were moved from India to the Far East command, that is to Malaya and Singapore, and two more were sent out from the United Kingdom. With the two torpedo squadrons and the two flying boat squadrons already at Singapore, this meant eight squadrons on the eastern flank.

Further west, 8B at Aden was strengthened by being given a fighter flight, with Hawker Demons. Late in 1939, this was replaced by 94F, new raised. Iraq proper was left at the outbreak of war with one bomber and one Army co-operation squadron, as well as flying boats at Basra.

There had been no fighter units overseas until 1936 when one day bomber squadron in Egypt (6B) was re-armed with Demons to replace its Harts. In 1938, 33B at Ismailia was similarly given Gladiators in place of Harts, and late in 1939 80F was sent out from the United Kingdom, being replaced by a new squadron at home. At the outbreak of war Egypt and Palestine, that is the Mediterranean seaboard, were defended by these three fighter squadrons, and by one Army co-operation, six bomber, and two bomber-transport squadrons. There was still the one seaplane squadron at Malta.

In other words, the overseas territories were not forgotten, but they had to wait till the home island was adequately defended. Internal security was run down in favour of external defence on the sea frontiers of the Empire.

Results

Now we can return to the question with which we started, how many sorties can we put up, if war started tomorrow? And suddenly, the answer is no longer as simple as it was in 1934, when we started. It turns into the questions whether the RAF could have gone to war at the Anschluss, or during the Munich Crisis, or at the final Czech crisis of March 1939. This may be an interesting question to the present-day historian or peace campaigner: in 1938, to the German Air Attaché, it was vitally interesting.

Of a sudden it is not enough to count pilots, or make estimates of how many of the pilots in post would have been fit to fly in action. In 1934 it could be assumed that each squadron had been together as a unit for some appreciable time, and only included a small number of learners, and even these could, under pressure, have been taken into the air to fight. But the squadrons of early 1938 were not merely twice as many as there had been in 1934, they were *all* of them new squadrons, with new and relatively inexperienced commanders, and a very green and raw middle layer of command in the flights. The new squadrons had been formed in many cases by splitting whole flights from old squadrons, and expanding them: and in some cases, splitting a flight from a new squadron itself just formed from a flight of an older one. The old squadrons of the original RAF had been split up and, for practical purposes, destroyed.

It may plausibly be claimed that the only useful squadrons left, keeping intact their old traditions, with their squadron and flight commanders still in place and with the habits of discipline firmly holding, were the Auxiliary squadrons, some re-titled from the old cadre squadrons. It might have been better in the end, if the money could be found, to have stuck to Sykes' plan. The other, Regular, squadrons were at least in 1938 still in such a state of confusion that they could not be seriously regarded as fit for war.

We have mentioned that the German Army as it went into Austria was not yet the well-oiled machine it was when it came against the French in 1914 or 1940. But nobody knew that at the time, and the inefficiency of the enemy is not a reliable guide for one's own standards. The Air Staff knew how confused the RAF was, and knew that this was an inevitable result of their expansion plan. With half their bomber and fighter pilots youngsters with less than a year's experience in squadron flying, any operations would have been token demonstrations, with heavy casualties to be expected against any opposition.

But at least March 1938 was better than March 1937, when even by the standards of 1938 the squadrons had been only half manned; the squadrons existed on the paper of the *Lists*, but in effect only there. By September 1938, the manning level was getting well up, but the pilots were still green, and inexperienced not merely in role and on type, but in the air. Even March 1939 was little better, but at least by now the problems of immediate supply of pilots were more or less solved.

Only solved numerically. Several pilots of the time complain of the rigid fighting instructions of Fighter Command, and of the defects of the current formations which were variants of the three man Vic, up to three or four Vics flying together.[11] It had been known since 1918 that the Pair or the Finger Four were superior formations for air fighting.[12] But the basic Pair depended on the protection of the more experienced leader, who was there to make the kills, by a less experienced wing man. As you can only learn to fly by flying, so you can only learn to fight by fighting. Any boxer will tell you that sparring is not enough. In 1939, the fighter pilots for the most part had not enough experience of flying to lead, and no experience of fighting at all, not even at second hand. The Germans had some corporate, if not always individual, experience of air war in Spain. Nobody yet knew who were going to make fighter leaders, and there were barely enough experienced flyers around to lead Vics of flights or squadrons. Even formation flying based on Vics was difficult enough and had to be learned.

The Fighting Instructions were rigid because the pilots were not yet good enough to follow any tactical instructions more sophisticated, not yet good enough to ad lib. They would be good enough as soon as they had enough flying, and a little fighting, in their log books. But not yet. They were still in the position of the infantry of the Somme, who were so inexperienced in war that all they could be trusted to do was to march forward in straight lines.

Where had the original squadron and flight commanders gone? In 1914 the General Staff had left their proper desks in London and gone off to command the Army: in a reverse move in 1937 and later, the squadron staffs had left their aeroplanes and gone off, not always willingly, to run the Air Force. It was, they knew, soon going to be a much bigger Air Force than anyone had yet imagined. Apart from all the problems of supply of a much larger number of stations and men, and the problems of recruiting and training the ground staff, were two vast areas for planning.

One area was the arranging and working up of the radar-based control system. It was being put in, physically, over the winter of 1938–9, and was ready for a large scale trial in the spring exercises of 1939.

The other was the re-arming of the squadrons. Up to 1937, the country had a biplane Air Force, except for a small number of Ansons, and the temporary expedient of such monoplanes as the British Bombay (50 built), Fairey Hendon (14 built) and Handley Page Harrow (100 built). During the following eighteen months, practically all the first line squadrons were re-equipped with monoplanes.

Re-arming came first to the bomber force. The Virginias and Sidestrands went, and were replaced by the Hampdens and the Whitleys, but in September 1938 there were still five squadrons of Heyford biplanes, as well as one of Hendons and five of Harrows, clumsy and outmoded monoplanes. The Hart developments were being replaced by the Battle, the Wellesley and the Blenheim, but there were still three Hind bomber squadrons on the *List*, and four of the Army co-operation squadrons were noted as having Hectors. Some of the general reconnaissance squadrons of Gauntlets, six of Gladiators, two squadrons of replaced Hinds. Fighter Command was the last force to change to the monoplanes. In September 1938, there were still eight squadrons of Gauntletsand, six of Gladiators, two squadrons of Furies, and three of Demons in the United Kingdom: four squadrons were re-equipping with Hurricanes and one with Spitfires.

With a side glance at the *Estimates*, we can now see how the manning plan and the task charts of the 1934 Air Staff were coming to fruition. The tactical structure of the RAF with its strictly functional system of Commands and Groups and the operational units, the squadrons, was in being by Munich time, and the airfields, if not absolutely finished, were ready enough. The manpower was recruited and was being groomed; it might take six months from wings to train a pilot to take his place in the line of battle, but it took eighteen months to make a night bomber captain, commanding a two pilot aeroplane. There would be enough reserves. The only thing to come, as already planned, were the aeroplanes, and the *Estimates* show that they were to be bought in 1938 and 1939. It would take the first months of 1939 to convert the crews and turn them into squadrons and re-equip even the Auxiliary squadrons with modern machines. The service was not yet ready to fight at Munich, but the planners had not meant it to be ready then. It would be ready by August 1939, when Kings in Europe gather in the barley harvest and take the field. There was yet to come an unexpected bonus: the 'Phoney War'. But from September 1938 on, Hobbes' dictum applied: Time was of the Nature of Warre.

Notes and References

1. M. Okumiya and J. Horikoshi, *Zero*, Cassell, 1957, Chapter 18. The whole book, although light in treatment and lacking in figures, is valuable as a sample of 'the other side of the hill.'
2. See Group Captain A.J. Capel, 'Air Co-operation with the Army', *RUSI Journal*, February 1939 for the doctrine at the outbreak of war. Compare *Flight*, 20 August 1933. At the Experimental Force exercises of 1927, Wing Commander Leigh-Mallory supported the tanks with low flying and strafing and got into trouble for diverging from this policy (B. Bond, *British Military Policy between the Wars*, p. 144).
3. Tedder in his autobiography claims responsibility for the introduction of civil flying schools. Higham (*Armed Forces in Peace Time*, Fowles 1961, p. 151) takes this to mean that Tedder at this point invented flying training schools, and that previously all pilot training was a squadron responsibility. In the German system, all pilots came through the Hitler Youth Glider Schools. They began powered training at the National Socialist Flying Schools, and after about sixty hours proceeded to the squadrons where they were brought up to 160 hours for qualification.
4. John W.R. Taylor, *CFS*, Putnam, 1958, Chapter 12.
5. A. Harries, *Bomber Offensive*, p. 25. See *The Aeroplane*, 19 April 1933, for an account of flying boat training.
6. *Flight*, 15 February 1933. In earlier years such an accident would have led to a courtmartial for the Squadron Commander (Sholto Douglas, *Years of Command*, Collins, 1966, Vol. I, p. 16).
7. B. Embry, *Mission Completed*, Methuen, 1957, p. 101. A scheme for training airmen Observers, who would work half their time as Observers and the other half at their trade, usually an Armament or Signals trade, was announced in *Flight*, 16 August 1934.
8. AP125, 1936, printing pp. 74, 155, 242.
9. Embry, op cit, p. 94.
10. *RUSI Journal*, January 1937.

11. J.E. Johnson, *Full Circle*, Chatto and Windus, 1964: R.P. Beaumont, 'Experiences in Military Flying during the Second World War', *Aeronautical Journal*, June 1972.

12. F.M. Green, 'The Development of the Fighting Aeroplane', *Aeronautical Journal*, February 1922, p. 46.

Chapter 13

Consequences and Truth

Beginnings and Endings

The RAF was constructed to make war. It began with Colonel Capper clinging to his man-carrying kite above Salisbury Plain, pulled by a team of horses harnessed to a GS wagon. Sykes worked out the plan, and Henderson pushed it through. In the middle of its first war, Henderson sat at Smuts' elbow and out of Lloyd George's political jobbery made a new Service which nobody then expected to last. Trenchard turned the squadrons in France into a human society, looking to him as their father figure, and was flexible enough to accept the new image of strategic air power when Weir put to him Sykes' basic idea and offered him the Independent Air Force. Through the bad days after the First World War, Trenchard built an Air Force on a new social pattern using a relatively new educational system, and kept it in being against Army and Navy opposition.

After Trenchard, his pupils, Ellington, Newall, Dowding and the Air Staff designed a new Air Force, using Trenchard's social basis in the squadrons, but made ready to fight one particular way, one campaign, even one battle, in the traditional way of a Great General Staff. They set out a Task Chart and a Manning Plan which imposed on them a date, spring 1939. They put through a monoplane revolution to re-shape the biplane squadrons, brought in the twin-engined medium bombers and ordered the four-engined heavy bombers for their preferred war date of 1941. They had designed a new pattern of ground maintenance shaped around Trenchard's Brats, and they had taken aboard Watson-Watt's radiolocation, made it work, and kept it a secret.

Now in September 1939, wearing the uniforms designed to fight the Boers in, committed to tactics worked out in the Air Exercises of 1933, the RAF made ready for a war in which their weapons would play a full part alongside the Army and the Navy. How full a part they did not yet realize.

Consequences

The Air Staff based their tactical plan on the only evidence they had, the Air Exercises. In strategic bombing, they would hold the North Sea and destroy the German Navy, flying great bomber sorties in the Box formation which would hold off any fighter attack. In the same way, they expected the Germans to attack in large bomber Boxes. The new radiolocation (not yet radar) warning system would allow the Civil Defence to expect the attacks, but there was little hope that the fighter squadrons would be able to break up the raids, let alone turn them back. Perhaps sufficient toll could be taken of the looser bomber formations on their way home to discourage them from returning. In any case, their own bomber attack, as unstoppable as the enemy's, might so disrupt the attacker's airfields as to block any further assaults.

The next events are well known, but their meaning is not always understood. During September to December 1939, a number of attacks were made on the German Navy, either at sea or in harbour at Brunsbüttel. Aircraft engaged were Blenheims, Hampdens and Wellingtons. Losses on these raids varied from one third to one half of the aircraft which actually found the targets.

The effect on the crews was immediate. Wellingtons had a metal skeleton covered by fabric. When they catch fire they burn very quickly. These attacks were made in daylight. 'You would sit there and watch your friends go "Whoosh" in flames, and then another "Whoosh".'[1] Decimation is often used as a synonym for destruction: it means a loss of 10 per cent. In war games it is usually taken as a very reasonable rule that a unit which suffers on this scale will not be fit for use again in that battle. During the attacks on Germany, Bomber Command suffered losses on this scale night after night, but continued the battle. Losses at five times this scale come nearer to total destruction.

The effect on the Air Staff was not so immediate, but just as real. The Box had been the basis of tactical planning, and of production. The events of the first months of the war showed that it was vulnerable to anti-aircraft fire and to fighter attacks when properly planned and directed. To abandon the daylight attacks meant jettisoning five years of planning. On the Bomber Command side, the decision was made in January to jettison, not only the Box, but the whole idea of strategic day bombing. The stereotype of the staff officer is not usually seen as having this degree of flexibility. 4 Group, armed with Whitleys, was already carrying out a night campaign over the German mainland, dropping leaflets. The experience was being built up in night flying, high flying, flying in extreme cold and bad weather. This was to be the new plan, the new system for the strategic bombers.

Other air forces were not so ready to drop the idea of day bombing by formations and of the defensive Box. The Germans tried it in 1940, and the Americans in 1942. The Germans found that it was only in the slightest satisfactory if they could provide fighter escorts, and not always then. The US Eighth Air Force found it prohibitively expensive, and only the introduction of the Mustang as a long-range escort enabled them to re-start day bombing after a hiatus.

Night bombing seemed to be a satisfactory mode of attack. By the end of 1940 the Operational Research Group at Bomber Command were able to show from photographs taken at the bomb release point, that very few of the night bombers were getting to their targets and most of the bombs were being dropped anything up to twenty miles or so away. The Royal Air Force was now reaping the reward of its long neglect of navigation. Useful bombing had to wait for the development of airborne radar.

It was the theory of the impenetrable Box which had prompted the design of the bombers, twin-engined or four-engined, with three or four turrets, guns, gunners and ammunition. Freeman Dyson has told us that on joining Bomber Command OR Branch in 1944 as a mathematician he was horrified on working out that the best way to reduce Bomber Command losses would have been to give the aeroplanes another thirty knots of speed and another 5,000 feet of ceiling to take them out of range of fighters and shell fire, and this would be possible by doing without the weight of the armament. Fewer aeroplanes would be lost, and those lost would each have three fewer men to be killed.

This was not original. OR Group folklore of the early 1950s was that the work had already been done by the Civil Service mathematician whose desk young Dyson took over, and who had now been sent to India to undertake the analysis of the bomber campaign carried out by Liberators flying from Bengal against Rangoon. By dispensing with the guns, gunners and ammunition, and by limiting the fuel load to what was absolutely necessary for the trip, the bomb load could be increased to something better than a token. But, when they were consulted, the crews preferred not to go thus unarmed, whatever the boffin said.

Japanese fighters had never been reported in Burma, and those that might be sent were of low performance, but, you never knew, they might appear. And the weather might be eminently predictable and the monsoon season might be over, but there could always be a freak storm, you never knew, or an airfield might go out of business, and a diversion might require more fuel. Besides, who ever heard of a pilot taking off without as much fuel as the aeroplane could carry, even if that meant topping up at the end of the runway before the take-off run. And most potent of all, *they* had designed and built the aeroplane with their armament, and *they* would never have done that if the weapons were not essential. Mosquitoes were different, *they* had *designed* the Mosquitoes without any guns, and *they* must know best. Native caution, stubborn conservatism, and blind faith in authority all coincided in their effect.

But the most important effect of the Bomber Command autumn campaign in the North Sea was not on the Bomber Staff, but on Fighter Command. The Box *could* be broken open. It *would* be profitable to attack the incoming raids, to break them up well before they reached their targets, perhaps even turn them back, without unacceptable losses in fighters. Not without losses, but with at least an exchange of losses, instead of a massacre. Fighter Command could be turned into a viable operation, rather than a gallant gesture. The effect was immediate. During November and December 1939, Fighter Command formed fifteen new squadrons, that is fifteen new units to be worked by the sector controllers, in the United Kingdom.

Thus within a couple of months of the outbreak of war, the fighter force within the United Kingdom was increased from twenty-five to forty squadrons. When Dowding was objecting to the detachment of more squadrons to France, he was still well up on the number he had originally thought to be sufficient. In any case, by the time the German Air Offensive opened, the detached squadrons had been brought back from France and Norway and there had been time to re-form them. The Battle of Britain was fought by a far larger British fighter force than had been planned. This was the effect of improved air traffic control methods.

In sense it is true to regard the Battle of Britain as fought by 11 Group. The attacks, after all, came in the 11 Group area at the centre of the L-shaped line of battle. But 10 Group (formed like 13 Group late in 1939) on the right and 12 Group on the left were not there simply to help 11 Group, although they did directly or indirectly as the shape of the battle demanded. They were there to hold their own sectors of the line.

The bomber losses of the opening months of the war may be said to have been avenged in the attack by Air Fleet 5 from Denmark across the North Sea against the East Coast on 15 August 1940. Caught far out over the water by the squadrons of 12 Group put there years before for just this purpose, they were slaughtered and did not try it again. The real question is, why did the Germans, knowing what they now did about the vulnerability of the bomber formations and about British speed and accuracy of response, try it the first time? The importance of this day's fighting was that the Germans *could* have tried it again, at any time, and it was a warning to the Air Staff that 12 Group must not be weakened unduly either in terms of squadrons or in terms of experienced pilots. This in chess terms was a highly successful gambit, the Germans gaining a striking strategic advantage at the expense of a material loss and pinning down half of Fighter Command.

It must be remembered that the greater part of the RAF in Britain in 1940 was not directly engaged in repelling the German air attack. The flying boats and the general reconnaissance squadrons went about their own task of looking for submarines. Mr Deighton makes much of the lack of hospitality afforded to a fighter squadron diverted to RAF Warmwell.[3] But Warmwell was not a Fighter Command Relief Landing Group, or an empty hostel for pilots. Warmwell had just opened as a Coastal Command Station, and its commander was trying, against the employment competition of defence plants at Weymouth and Holton Health for civilian catering workers, to get his unit supplied and fed and to keep his two Anson squadrons flying on a continuous sweep of the Western Approaches, newly left open to the Germans. Warmwell was fighting the Battle of the Atlantic, noting with interest some fighter activity nearby.

And the bomber squadrons? An Air Marshal who had better remain nameless told the author: 'There was no fighter Battle of Britain. I was at Lympne in light bombers in 1940. There was some fighter activity overhead but no more than you would expect. We went out every night, destroying the German invasion barges in the Channel Ports. That was why the Germans never came. *We* fought the real Battle of Britain.' The argument is still open.

The Sea War
The end of the naval story came in June 1940. When the Germans invaded Norway, and Allied armies were put in at Narvik to repel them, two carriers, *Glorious* and *Ark Royal*, went with the covering naval forces. The carrier-borne fighters proved no match for the German land-based fighters. And whatever the quality of the carrier aircraft, there were simply not enough of them.

Two squadrons of RAF fighters, 46F with Hurricanes and 206F with Gladiators, were brought to Norway in *Glorious* to operate from land bases, in one case from a frozen lake. When the Army was withdrawn in June 1940, it was proposed to bring out the men but abandon the aeroplanes. This was not to the taste of the RAF. The Navy had based its whole system on the theory that only divine beings spurning the solid earth and walking like Gods among the clouds, in other words real naval officers in dark blue double-breasted uniforms, could possibly learn, and then only after much skilled instruction, to land on a carrier. The two senior RAF officers present, Wing Commander R.T. Atcherley and Squadron Leader K.B. Cross, thought otherwise, and the normal run-of-the-mill squadron pilots of the RAF landed ten Hurricanes and seven Gladiators (the unmodified land variety) on *Glorious* without accident or incident. The Captain of *Glorious* then decided to go home without the rest of the fleet to arrange a courtmartial quite unconnected with this campaign. The Navy with some precision vectored *Glorious* to an interception course with *Scharnhorst* and *Gneisenau*. So all the precious aeroplanes were lost and almost all the brave and resourceful men who had tried to bring them home were drowned.

Revenge had to wait, although the two destroyers escorting *Glorious* attacked, were sunk with only one survivor each, and put one torpedo between them into *Scharnhorst* which sent her into dry dock for the rest of the summer. But the two battle cruisers and the cruiser *Prinz Eugen* were got into Brest during the summer of 1941, where the RAF attacked them night after night. Eventually in spring 1942 the German High Command were forced by the bombing to take the dreadful risk of withdrawing these heavy ships from that position of great tactical advantage and bringing them home by a run through the Channel. They got past Dover, and this successful gambit by the RAF and the Royal Navy was proclaimed by the media on both sides as a great German success and as a humiliation for the British.

But all three ships fell victim to the minefields already laid by the RAF against just such an eventuality. *Gneisenau* went into dry dock in Kiel and was so damaged there by later air raids that she was virtually abandoned. *Prinz Eugen* was repaired in time to be surrendered to the Americans and used as a target at Bikini. *Scharnhorst*, like *Tirpitz*, led a wandering life hiding from the bombers among the Norwegian fiords till she was caught in the North Sea and sunk by *Duke of York* on Boxing Day 1943.

But all this was to come. Bomber Command went on with learning how to operate in the dark, and the squadrons in France went on with learning how to operate as equal partners with the Army. Meanwhile the German General Staff were busy at their drawing boards. The old plan would do, mostly. They already occupied the airfields around Prague which could threaten

Berlin. They would secure their right flank by first attacking Norway and Denmark. In the main assault they would this time take in Holland, and move quickly into Antwerp as they went through Belgium to the north of the impregnable French frontier of the Maginot Line. And also to the north of the impassable and therefore unfortified Ardennes. As they had the spare capacity, they would attempt the passage of those impassable hills and forests with two divisions, one commanded by Erwin Rommel, and both armed with the Skoda tanks which the year before had been the first line equipment of the Czech Army. Meanwhile although British squadrons would need to be dealt with in the north, the French Air Force was divided by the need to repel potential attacks from Italy and Spain, for while the armies must come through the passes, the air forces could come in by any way they liked. Franco knew there were safer ways of helping in a war than fighting in it.

Twenty-two years after it had been formally founded, the RAF again sat in the sunshine of a spring waiting for the assault they knew would come.

Questions and Answers

This account of the way the RAF was prepared, and prepared itself, for the great assault, leaves a shoal of questions, some unanswered.

The first is quite straightforward. Why does it appear that conventional historians do not use the mass of numerical material which can be derived from such a good source as the *Air Force Lists*? This is material which has not been subjected to any later treatment. It has not been altered by adjustment to suit any fixed opinions, any wish of the Establishment to convey this or that impression. It may be that any historian who uses numerical material, other than financial or trade statistics, runs the danger of being accused of the crime of cliometry. Probably academic students of history, young research students in particular, are simply not taught to count, and are more afraid of adding up columns of figures than of the eye-breaking labour of reading official documents of any century. The drudgery could be done on a computer, if only anyone had the patience, or the labour and the money, to put the old *Lists* in a usable form on to the computer file. But computers are only useful if the researcher knows what questions to ask. Most of the work here was done by hand and eye, because as so often it appeared that even if the source documents could have been abstracted for recording, once the material had been arranged for the programmer to deal with, most of the additions and distributions would have been done anyway. Only immersion in the raw data will tell you what computer questions to ask.

Related to this is the other question: why did no one use this material at the time it was issued? That it was not used is shown by the number and nature of the questions asked in Parliament during the re-armament period. Members seem only to have been interested in the gross number of aeroplanes available on one side or the other. Sometimes the answers given are suspect. If the numbers of operational or front line aeroplanes were asked for, it seems that the PQ was passed down in the Ministry to someone who counted up the number of squadrons authorized or existing, and from that went to the number of aeroplanes on their several establishments. Whether the aeroplanes were there or not, whether they were in themselves of any use, and whether there were pilots available to fly them or not, did not necessarily enter into the calculation.[5]

Perhaps the old stores device of conversion was sometimes found of use; in preparing an inventory of, say, a blanket store for someone else to take over, it is found useful to convert a deficiency of five serviceable blankets into a surplus of one damaged blanket by the use of brute force and a considerable noise of tearing, as well as the request for the replacement of, in total, six damaged blankets. There are other useful variants of conversion, largely based on verbal similarities, although it is usually thought unfair to convert hangers coat into hangars aircraft.

Now, you will only understand a process of this kind if you have yourself been once persuaded to take over the blanket inventory. But many of the MPs of the interwar period had been at some time or other service officers, not simply wartime officers but often regular officers, and would be familiar with these processes of military imaginative accounting and also with the *Lists*: this is especially true of the Conservative and Liberal benches, and at least of Major Attlee. One can understand that non-military politicians like MacDonald, Baldwin or

Chamberlain might take what they were told by the Ministries as true, but the fact that the rank and file of back-bench MPs (from Churchill, late a Hussar officer, down) found it necessary to ask officially, or unofficially to beg their friends who were generals, for information they might easily have found out for themselves from the *Lists* and *Estimates* does not lead us to rate their insight or their diligence very highly.

But while the politicians obviously did not use the *Lists* in the libraries of either House, or like anyone else buy them in the Government Bookshop, others were interested. It is likely that the *Lists* were issued free monthly not only to military stations and to other Government Departments, but also to Embassies in London for the use of their Air Attachés. Not that it mattered whether it was issued free or whether the attaché sent his clerk up to Kingsway to buy it. The information on the manning of the RAF was available in considerable detail to all and sundry. The Germans could plot the growth of the Air Arm, with all its stations neatly indicated on the map, with all the units shown with their operating strengths in pilots and the types of aeroplanes.

But did the Germans bother? There is no very obvious evidence that they had this knowledge of the strength and location of the RAF. On the raid of 15 August 1940, mentioned above, it would seem that the German squadrons launched from Denmark were surprised to find the British waiting for them in force. The Order of Battle was as little known to them as was the existence of radar and the meaning of the radar stations, and they really *were* kept a secret.[6] The Germans knew the tall aerials were an essential part of the British defence system, but not quite how essential, and oddly enough neither the *Lists* nor the *Estimates* told them anything. Our interpretation of the Germans' political thinking before the opening of the war and of their movements during the Battles of France and of Britain would be very different if we knew on exactly what information they were basing their decisions. What in fact we want to know is whether Göring ever heard what his Air Attaché was reporting.

There was a defect in the Air Staff planning. They thought that bombing German industry, set in the German cities, would win the war. But in Britain, in 1936, someone who knew a bit of history, especially Spanish history, remembered Saragossa and looked at Madrid, where it wasn't simply a matter of aerial bombing but of the 88s firing over the parapet at the end of the street. He realized that what might be true of a third world city under fire might not be true of a Western industrial city where the people were politically united, literate, technically competent and resourceful – and organized. So Civil Defence was born. Surely the air marshals must have realized that what we had done, the Germans could do, and bombing might not work, might not even reduce output?

Another question to be discussed is one of ethics. It is clear from the figures that the Air Staff planned the Battle of Britain, from late 1933. It was clear to the thinking officers who had read *Mein Kampf*, or at least been briefed on its contents, that the Germans intended to fight someone in a few years and the British were as good an enemy as anybody: the political instability of France did not encourage reliance on that ally. It was clear to every schoolchild who listened to Stephen King-Hall on the Radio. It seems to have been clear to most people except for the politicians who were supposed to make decisions and to plan. The Air Staff looked ahead and planned, and merit praise for it. Nobody else deserves much credit. We know precisely who these Air Staff officers were. Air Marshal Dowding was, in 1933, Air Member for Supply and Research: the key directors in the various planning sections were Group Captains Arthur Harris, Peck, Evill, Garrod, Edmonds and Tedder.

If you plan for a war, however inevitable it may appear, are you not in some way one of the causes of the war? If so, then have you done wrong morally? The war of 1939 was without doubt a Just War, against an evil enemy. To a soldier, of course, any war must seem a Just War, for the defence of his country. No thinking person in the period 1933–39 would have called the war of 1914 a Just War, but it was not immediately obvious in 1939 how just a war this new conflict was to be. The truth was not to be seen fully till the concentration camps were liberated.

But if the war were just, then surely it should have been opened as soon as the evil was recognized. Any delay cost the lives of the oppressed. But the plan worked out so carefully in 1931 called for delay till 1939 at the earliest, or the defence would not be effective. Britain was not ready to fight at Munich because the Air Staff had planned it that way. The politicians and civil servants might produce new lettered plans, but except as an indication of their good intentions they were hardly worth the paper they were torn up on. Once Plan 'A' had been approved and the staff went to work on implementing that Task Chart and Manning Plan, there was no hope of stopping it, any more than there had been any hope of stopping German mobilization in 1914 once the trains had begun to roll. Or of speeding it up. Time was of the nature of War. There were vast tasks of airfield building, of training the technical workers, of training the pilots, and of making the aeroplanes. All had to be co-ordinated in infinite detail. Were the Air Staff wrong in saying, very firmly, that they had no intention of going to a war that they could not hope to win?

There is a related question. I have argued that it seems that the Air Staff did not so much deceive the Ministers as economize drastically with the truth they doled out, agreeing soothingly with all the successively lettered plans produced in one panic after another, but taking no real notice while they went ahead to produce a workable and formidable Air Force. Is a General Staff in a democratic state ever justified in going so far in using its professional knowledge and its initiative independent of the politicians? It cannot be called treason because it prospered. But if it had not succeeded?

Last we may ask who fought this Battle. A poet of the time, Francis Brett Young, gave the conventional answer:

> Who were these paladins? ... They were the seed
> of the mild unassuming Middle Class,
> The sons of lawyers, parsons, doctors, bankers.
> Shopkeepers, merchants, chemists, engineers.

This is certainly true: the NCO pilots, the common soldiers, need not be ignored, they were certainly in social and financial origin a few steps above the Other Ranks of the traditional Army. Young was a country doctor who had made a fortune out of writing, and his biography, put together from his notes by his wife after his death, shows that he had come to mix with the landed gentry and to patronize the class above which he had now risen. What he was saying, although he probably was not sophisticated enough to realize it, was that these paladins came from the upper 10 per cent or fewer of the nation in terms of income: the pilots were the upper classes, the nobs, the toffs, the rich, the well off, them above, the bosses, put it how you will. As far back as the Roman Republic, it was a privilege of the wealthy to fight for the State in the most interesting and exciting way. The pilots of 1939 came typically from the public schools. The contribution of the mass of the population, of the grammar schools, was yet to come, in the days of the Bomber Offensive: it came largely because in 1940 someone dusted off the Sykes Memorandum and activated a cadet force for boys, the Air Training Corps.

But does it matter who they were, wealthy or not? They *were* Paladins, they were volunteers, their shoulders held the sky suspended. And they saved us from the jackboot.

Notes and References

1. Wing Commander Perioli, personal communication.
2. Freeman Dyson, *Disturbing the Universe*, Harper and Rowe, 1980. OR Branch folklore, communicated by F.C. Watts and others.
3. Len Deighton, *Fighter*, Jonathan Cape, 1977, part 4. In this book, Mr Deighton sees the Battle of Britain as mainly a matter of machinery. In his novel *Bomber*, he sees the Bomber Offensive as an incident in the class struggle: he may well be right.
4. John Winton, *Carrier Glorious*, Leo Cooper, 1986, washes a good deal of naval dirty linen.
5. J.F. Kennedy, *While England Slept*, Hutchinson, 1940, p. 112, notes that Churchill was interested only in gross numbers of squadrons and aeroplanes without going into details.
6. 'In May 1939 I was posted from command of Oxford University Air Squadron to be a Controller at HQ Fighter Command at Bentley Priory. During the summer exercises of 1939 we were taught how to work from the big plotting table. We were told it was no business of ours how the plots were produced to be put on the table: we were to treat them as reliable. In August 1939 I was posted to command a fighter squadron. I was then taken through the green door and shown the filter plot, where the raw data came in and was tidied up to be fed to the plotting table. I was told it was produced by wireless, and nothing more.' The late Air Vice-Marshal James Kirkpatrick, personal communication.
7. Francis Brett Young, *The Island*, Heinemann, 1944, pp. 449–50. Charlemagne's Paladins, riding against the heathen, were the great landowners of his empire.

Tables

1. *Army List*, August, 1914.
2. Disposition of RFC and RNAS squadrons, March 1918.
3. *Navy List*, April 1918.
4. *Air Force Lists*, 1919 and 1920.
5. *Air Estimates*, by years.
6. *Hardware Estimates*, by years.
7. RAF manpower by years.
8. Growth of the Ground Branches.
9. Location of Squadrons by selected dates.
10. Pilot Strengths in the UK, by years.
11. Naval Aviation: list of carriers.
12. The Expansion Plans.
13. Flying Stations in 1933 by formation.
14. Stations by date of formation, 1934-9.
15. UK Squadron formation, 1934-9.
16. Interwar bomber data.
17. RAF organization, September 1938.
18. The Group Pool Squadrons.
19. Airman Trade Groups and pay, 1919.
20. Airman Trade Groups and pay, 1934.
21. Officer pay scales, 1934.
22. Civil Service pay scales, 1934.
23. RAE Farnborough pay scales and manning, 1934.
24. Officer and aircrew pay scales, 1938.
25. Civil Service scientific and technical pay scales, 1938.
26. Civilian strengths RAE, 1938-9.

Table 1

Army List, August 1914
Royal Flying Corps

A: Military Wing

Officer Commanding, Temporary Lt.-Colonel Sykes (Substantive Brevet-Major)

Adjutant, Officers in charge of Stores and of Workshops (all graded as Flight Commanders), Quartermaster

Squadron	Station	Squadron C'mdr	Flight C'mdr	Flying Officers
1	Farnborough	1	2	
2	Montrose	1	3	10
3	Salisbury Plain	1	3	12
4	Salisbury Plain	1	4	11
5	Farnborough	1	2	12
6	Farnborough	1	2	8
7	Farnborough	1	1	1
8	Not yet formed			

		Squadron C'mdrs	Flight C'mdrs	Flying Officers
	CFS	1	3	
	Aircraft Park	1		1
	Airships		1	5
	Kite School		1	5
	Total	10	22	63

RFC Reserve: 32. All rated as 'Flying Officers', regardless of substantive rank, including Colonel (Temporary Brigadier-General) Henderson.
RFC Special Reserve: 20. All 'Second Lieutenants on probation'.

Note: Squadron Commanders are substantive Majors or Captains, Flight Commanders are Captains or Lieutenants, Flying Officers with the squadrons are either Lieutenants or Second Lieutenants.

B: Naval Wing

Central Flying School

Commandant: Captain G.M. Payne
Assistant Commandant: Major H.M. Trenchard CB, DSO
(graded as Squadron Commander)

Naval Wing units:

Central Air Office
Naval Flying School, Eastchurch
Isle of Grain Naval Air Station
Calshot Naval Air Station
Felixstowe Naval Air Station
Yarmouth Naval Air Station
Fort George Naval Air Station
Dundee Naval Air Station
Kingsnorth Naval Air Station

Farnborough Airship Station housing
Naval Airship No. 3, Lt. W.C. Hicks, RN (Flying Officer)
Naval Airship No. 4, Lt.-Cmdr F.L.M. Boothby RN in Command
(Squadron Commander)

Squadron Commanders	9
Flight Commanders	4
Flying Officers	39
Flying Officers u.t.	5
Flying Officers (reserve)	3
Surgeons	3

Warrant Officers:

Carpenters	3
Artificer Engineers	2
Gunners	3
Boatswain	1

H.M. Trenchard, CB, DSO, appears in this list as a Substantive Captain of seniority 28 February 1900, Brevet-Major of seniority 22 August 1902, in the Royal Scots Fusiliers.

Other Trenchards in the list include O.H.B. Trenchard, Captain R.E., F.A. Trenchard, Lieutenant R.A., and G.B.B. Trenchard, Lieutenant R.G.A. This contrasts with Trenchard's insistence on his uniqueness in name as described by Boyle, p. 27.

Frederick Sykes is shown as psc, Interpreter (French), Interpreter (German), substantive Captain 15th Hussars, seniority 1 October 1908, Brevet-Major seniority 3 June 1913, Temporary Colonel while in Command of the Military Wing.

Table 2

Disposition of RFC and RNAS Operational Squadrons by Theatre, March 1918

Theatre	RFC	RNAS
UK	37	31
Ireland	2	
France	60	9
		4 (attached RFC)
Italy	4	
Greece	1	
Egypt	3	
Palestine	3	
Mesopotamia	4	
India	2	
East Mediterranean		5
Independent AF	2	1

This table illustrates clearly that France was the main war theatre, the others being merely sideshows in which it was not thought necessary to incur the enormous expense of supporting a large air force, however much it might be helpful.

Table 3

Navy List, April 1918

The Royal Flying Corps
Colonel in Chief HM the King

(During the war, the List of the Military Wing is omitted)

Commodore: Sir Geoffrey Paine, Master General of Personnel, The Air Council

Wing Captains   14
(Ranks from Lieutenant
Commander to Captain, with
one Major Royal Marines, and
one Lieutenant Colonel
Essex Regiment)

Royal Navy

Wing Commanders	49
Squadron Commanders	109
Squadron Observers	4
Flight Commanders	197
Flight Observers	6
Flight Lieutenants	629
Observer Lieutenants	48
Flight Sub-Lieutenants	1,020
Observer Sub-Lieutenants	125

Royal Naval Air Service

Temporary Commanders	8
Temporary Lieutenant-Commanders	138
Temporary Lieutenants	747
Temporary Honorary Lieutenant	1
Temporary Sub-Lieutenants	194
Probationary Flight Officers	1,449
Probationary Observer Officers	125
Warrant Officers I	38
Warrant Officers II	487

The Air Branch is to be removed from the July List

This table shows the size of the Naval Air Effort and the origin of the later RAF Rank Titles.

Table 4a

Air Force List, February 1919

Major Generals	11
Colonels	35
Acting Brigadier-Generals	29
Lieutenant-Colonels	254
Majors	770
Captains	3,568
Lieutenants	11,914
Second Lieutenants	15,573
Total	32,129

Demobilization had not yet begun, and this probably represents the peak strength of the RAF in the First World War.

Table 4b

Air Force List, January 1920

Employed	
Air Officers	20
Group Captains	24
Wing Commanders	103
Squadron Leaders	320
Flight Lieutenants	1,212
Flying and Observer Officers	3,252
Pilot Officers	2,101
Total employed	7,032
Unemployed	
Colonels	1
Lieutenant-Colonels	60
Majors	375
Captains	2,205
Lieutenants	8,835
Second Lieutenants	11,173
Total unemployed	22,649

The Employed Officers included all Medical Officers of Wing Commander and above, but there were also 192 more junior Medical Officers on a separate list. There were three Medical Administrators, and twenty-eight Dentists. These numbers reflect the slow reduction of the RAF to a practicable peacetime size. We must assume that all the unemployed officers, as well as about half those still employed, were in the queue for demobilization. These figures should be compared with those given above for 1919.

Table 5

Air Estimates

Year	Total Estimate £k	Vote 1 £k	Vote 3 £k	Vote 8 £k
1920	21,471	4,661	2,353	895
1921	19,007	3,566	3,758	880

Table 5 contd.

Year	Total Estimate £k	Vote 1 £k	Vote 3 £k	Vote 8 £k
1922	15,667	3,781	1,295	411
1923	18,605	3,500	3,870	287
1924	14,861	2,941	4,029	355
1925	15,513	3,412	5,650	357
1926	16,000	3,405	6,091	462
1927	15,800	3,160	6,424	464
1928	16,563	3,401	6,567	415
1929	16,200	6,585	6,585	450
1930	18,177	3,731	7,596	500
1931	18,100	3,907	7,672	470
1932	17,400	3,930	7,350	473
1933	17,421	4,110	5,935	490
1934	17,561	4,210	7,220	513
1935	20,650	4,547	8,200	595
1936	39,000	6,518	18,491	760
1937	51,147	8,466	31,542	2,315
1938	73,500	10,175	42,621	2,925
1939	166,561	19,690	111,110	4,787

Notes: (1) Vote 1 covers pay and allowances, Vote 3 covers Warlike Stores, i.e. airframes, engines and similar hardware, Vote 8 covers expenditure on civil aviation.

(2) These estimates are net, and would in almost every case be increased during the year by supplementary estimates, grants in aid and other financial adjustments, well publicized or not. They do reflect, however, the relative importance attached in successive years by the legislators to aviation, civil and military, and to men and to machines.

Table 6

Hardware Estimates

	£k		
Year	Aeroplanes	Engines	Armament
1922	390	87	15
1923	1,913	941	40
1924	2,789	1,451	119

Table 6 contd.

Year	£k Aeroplanes	Engines	Armament
1925	2,908	1,537	111
1926	2,888	1,031	93
1927	2,930	1,574	61
1928	2,955	1,533	68
1929	3,370	1,910	75
1930	3,642	1,946	86
1931	3,448	1,706	112
1932	2,657	1,741	115
1933	2,804	1,749	103
1934	2,810	1,940	114
1935	3,577	2,246	130
1936	8,335	4,525	776
1937	19,550	9,450	1,740
1938	42,680	24,130	
1939	93,640		

Notes: in 1938 and later the details of this Vote were omitted in the Estimates. The armament sums are incorporated into the Aeroplane and Engine votes.

Table 7

RAF Authorized Manpower by year

	Officers Air	Other	WOs	NCOs	Airmen
1920	19	3,040	324	2,900	19,760
1921	19	2,913	329	3,433	21,845
1922	40	3,045	300	3,800	22,195
1923	31	3,186	250	4,124	22,731
1924	35	3,325	309	4,881	23,410
1925	35	3,592	330	5,360	24,500
1926	35	3,512	345	4,780	22,900
1927	35	3,401	320	4,700	20,525
1928	35	3,395	318	5,000	20,000
1929	38	3,300	420	5,000	19,880
1930	39	3,300	460	5,320	19,078
1931	38	3,200	500	5,500	19,432
1932	38	3,200	512	6,200	19,174

Table 7 contd.

	Officer		WO's	NCO's	Airmen
	Air	Other			
1933	37	3,158	550	6,070	19,042
1934	41	3,150	450	6,450	19,049
1935	43	3,255	443	7,011	19,096
1936	54	4,236	553	10,508	29,529
1937	67	5,247	600	13,553	42,800
1938	68	6,943	721	15,095	51,696

These figures represent not strengths but the number of men in the various ranks whose pay, etc, was voted by Parliament. They do not include the men of all ranks serving in India, since they were paid for separately by the Government of India.

Table 8

Growth of the Ground Branches

Year	Equipment	Accountant
1922	167	78
1925	260	117
1930	303	150
1934	245	152
1935	279	152
1937	394	169
1938	353	190
1939	420	201

The Equipment Branch started as the Stores Branch, and the Accountant Branch as the Stores Accountant Branch. The slow growth in numbers during the expansion period should be noted, and is explained in the text as due to the increase in the number of Stations rather than to an increase in manpower.

Table 9

Location of Squadrons by selected dates

A. Overseas

Theatre	Malta	Egypt	Iraq Aden	T/Jordan Palestine	India	Far East
1922		3	6			
1925	1	2	8	2	8	
1927	1	3	6	2	6	
1929	1	4	6	1	8	1
1931	1	4	6	2	8	2
1933	1	4	6	2	8	2
1935	1	4	6	2	8	3
1937	1	5	6	3	8	4
1938	1	5	6	3	8	4

This table illustrates the stability of British Air Strength in the overseas territories up to 1938.

B. United Kingdom: Regular Squadrons by type and year

	F	B	A/Coop	TB	FB	Com	GR
1921	2	2	2	2	3		
1925	9	9	4		1		
1927	12	9	4		1	1	
1929	12	10	6	1	2	1	
1931	14	11	5	1	3	1	
1933	13	12	5	2	4	1	
1935	14	15	5	1	5	1	
1936	16	24	5	1	5	1	1
1937	25	47	5	2	7	1	6
1938	25	56	7	2	7	1	6

Note: the two notional bomber squadrons at Martlesham Heath are included as long as they lasted.

Table 10a

UK Regular Fighter Squadrons: Pilot Strength by ranks and year

Year	S/L	F/L	F/O	P/O	A/PO	NCO	Total
1922	2	7	40				49
1925	9	19	52	46		15	141
1927	12	27	75	54		45	213
1929	12	28	60	56		59	215
1931	14	29	70	55		65	233
1933	12	39	73	19		74	217
1935	14	15	33	44	11	78	184
1936	10	23	24	55	31	66	209
1937	23	13	19	120	32	91	268
1938	25	18	45	240	32	159	437

Table 10b

UK Regular Bomber Squadrons: Pilot Strength by rank and year

Year	W/C	S/L	F/L	F/O	P/O	A/PO	NCO	Total
1922		2	8	37				47
1925		11	32	84	18		30	175
1927	3	13	22	58	51		77	230
1929	4	17	23	64	25		45	178
1931	5	17	31	77	63		91	284
1933	5	15	37	70	39		91	257
1935	5	21	45	39	61	14	89	272
1936	6	12	28	56	60	35	83	280
1937	12	50	29	44	325	83	183	731
1938	20	72	36	88	658	59	394	1,327

These two tables show how the operational Bomber and Fighter Squadrons changed dramatically in 1937 and 1938 from relatively stable structures to organizations short of middle management and massively made up of green pilots.

Table 10c

Pilot Strength in the UK

	Sqns	1st Line			Immediate Reserve		
		Officers	NCOs	Total	Officers	NCOs	Total
1922	11	246		246	362		362
1925	22	379	45	424	337	11	348
1927	25	380	122	502	364	36	400
1929	31	432	110	542	451	39	490
1931	34	415	166	581	539	30	569
1933	36	484	184	668	543	43	586
1935	41	451	176	627	546	33	578
1937	92	1,009	359	1,368	698	128	826
1938	103	1,487	705	2,192	643	186	829

First Line and Immediate Reserve are calculated as shown in the text. The relationship between the increase in squadron and pilot numbers should be noted. Up to the beginning of the expansion, the Service could call on over 70 per cent immediate reserve of seasoned pilots: this reserve was proportionately reduced as expansion got under way.

Table 11

Naval Aviation

Carriers

Ship	Date in Service	Complement		Speed Knots
		Aircraft	Men	
Ark Royal (UK)	1914	7	180	11
Furious (1) (UK)	1917	8	880	31
Argus (UK)	1918	20	401	20
Vindictive (UK)	1918	6	560	30
Eagle (UK)	1920	21	950	22
Hosho (Jap)	1922	26	550	14
Langley (US)	1922	6	468	15
Hermes (UK)	1924	20	664	25
Furious (2) (UK)	1925	36	1,218	31
Lexington (US)	1925	63	2,327	33
Saratoga (US)	1926	63	2,327	33
Bearn (France)	1927	40	875	21
Akagi (Jap)	1927	91	2,000	31
Courageous (UK)	1928	48	1,216	30

Table 11 contd.

Ship	Date in Service	Complement Aircraft	Men	Speed Knots
Kaga (Jap)	1928	60	2,000	28
Glorious (UK)	1930	48	1,216	30
Ryujo (Jap)	1933	48	600	29
Ranger (US)	1934	76	1,788	29
Yorktown (US)	1937	96	2,175	32
Soryu (Jap)	1937	71	1,101	34
Ark Royal (UK)	1938	60	1,580	31
Enterprise (UK)	1938	96	2,175	32
Hiryu (Jap)	1938	73	1,101	34

The most important figures here are in the Speed column. A carrier with a capability of less than about thirty knots would not be able to take her place in a battle fleet (maximum about twenty-three knots in 1939) or in a cruiser task force, because after flying off aircraft she would not be able to catch up.

Table 12

The Expansion Plans

Plan	Date	Output Date	Squadrons	Aircraft
A	11/33	3/39	76	960 (336F)
C	3/35	3/37	123	840 (420)
F	11/35	3/39	124	1,020 (420F)
H	1/37	3/39	145	1,631 (476)
J	10/37	6/41	158	1,442 (532)
K	1/38	3/41	145	1,360 (539)
L	5/38	3/41	141	1,352 (608)
M	10/38	3/42	145	1,360 (800)

As explained in the text, this table is of most use as expressing the varying states of good will towards the Air Arm by the Administration at different dates and under differing degrees of political pressure. Only Plan A, the first devised by the Air Staff, seems to have had any reality. Once the Service had begun to fulfil it in 1933, there was no more hope of altering it either in shape or in speed of development than there was in trying to alter German mobilization in 1914 once the trains had begun to roll.

Table 13

Flying Stations 1933, by formation

Station	Squadron
1 Group	
Abbotsinch	602B
Aldergrove	502B
Castle Bromwich	605B
Filton	501B
Hendon	600B, 601B, 604B
Hucknall	504B
Thornaby	608B
Turnhouse	603B
Usworth	607B
Waddington	503B
Wessex Bombing Area	
Andover	12B, 101B, Abingdon
Abingdon	40B
Bicester	73B
Bircham Newton	35B, 207B
Boscombe Down	10B
Upper Heyford	18B, 57B, 99B
Worthy Down	7B, 9B, 58B
Fighting Area	
Biggin Hill	23F, 32F
Duxford	19F
Hawkinge	25F
Hornchurch	54F, 111F
Northolt	24C, 41F
North Weald	29F
Tangmere	1F, 43F
Upavon	3F, 17F
Coastal Area	
Calshot	201FB
Donnibristle	100TB
Mount Batten	204FB, 209FB
Pembroke Dock	210FB
21 Group	
Martlesham Heath	15B, 22B

Table 13 contd.

Station	Squadron
22 Group	
Catterick	26AC
Farnborough	4AC
Manston	2AC, 500B
Netheravon	13AC
Old Sarum	16AC
23 Group	
Digby	2FTS
Grantham	3FTS
Sealand	5FTS
Others	
Cranwell	RAF College
Eastchurch	School of Anti-aircraft co-operation
Gosport	RAF Base
Leuchars	RAF Training Base
Wittering	Central Flying School

Note that although Martlesham has its two notional Bomber Squadrons lists, it is not a station in either of the two Bomber Groups.

Table 14

Stations by year of first approval and original intention, with September 1939 occupancy

Year	Station	Intention	Occupancy
1934	Mildenhall	2 Sqns	99B, 149B
1935	Feltwell	2 Sqns	37B, 214B
	Harwell	3 Sqns	62B, 82B
	Marham	2 Sqns	38B, 107B
	Odiham	2 Sqns	13AC, 53AC
	Stradishall	2 Sqns	148B
	Ternhill	FTS	
	Hullavington	FTS	
	Waddington	2 Sqns	44B, 110B
	Yorkshire	2 Sqns	(Linton on Ouse?)
1936	Bawdesey	Research Establishment	
	Church Fenton	2 Sqns	64F, 72F

Table 14 contd.

Year	Station	Intention	
	Cranfield	3 Sqns	62B, 82B
	Debden	3 Sqns	25F, 85F, 87F
	Dishforth	2 Sqns	100B, 78B
	Finningley	2 Sqns	7B, 76B
	Leconfield	2 Sqns	97B, 116B
	Little Rissington	FTS	
	Penrhos	Armament Training Camp	
	Scampton	2 Sqns	49B, 83B
	South Cerney	FTS	
	Thorney Island	5 Sqns and Training Flight	22TB, 42TB
	Upwood	3 Sqns	52B
	West Freugh	Armament Training Camp	
	Wytton	3 Sqns	114B
1937	Benson	2 Sqns	
	Bradwell	Air Firing Range	
	E. Anglia (Ely?)	Hospital	
	Evanton	Air Firing Camp	
	Honington	2 Sqns	75B, 215B
	Scilly Isles	WT Station	
	Wattisham	2 Sqns	
	Watton	2 Sqns	
	West Raynham	2 Sqns	
	Worcestershire	Equipment Depot	
1938	Acklington	Armament Training Camp	
	Brize Norton	FTS	
	Crichel Down	Bombing Range	
	Detling	1 Auxiliary Sqn	
	Manby	School	
	Midlands	Equipment Unit	
	Pembrey	Hutted accommodation range	
	Porthcawl (Stormy Down)	Armament Training Camp	
	Saint Athan	School of Technical Training	
	Sealand	Aircraft Depot	

Note that all the Armament Training Camps were originally approved as hutted camps, intended for temporary occupation by squadrons using ranges. At this period, all runways were grass, the Air Staff delaying the introduction of the heavy bombers as long as possible to avoid the expense of building concrete runways.

Table 15

Formation of Regular RAF Squadrons by type, and by year and quarter from 1934 to 1939, in the UK

Year	Quarter	Fighter	Bomber	Army Co-op	Reconnaissance
1934	I				
	II		2		
	III	1			
	IV				
1935	I	1			
	II				
	III		3		
	IV	2	3		
1936	I	1	6		
	II				1
	III	3	2		1
	IV		3		1
1937	I	7	10		
	II		10	3	1
	III	1			
	IV				
1938	I		1		
	II	1	2		
	III				
	IV				
1939	I			1	
	II				
	III			1	
	IV	15			

It can be seen that Squadron formation virtually ceased after the middle of 1937. Squadron formation continued after the outbreak of war in a sporadic dribble, except for the expansion of Fighter squadrons in late 1939.

Table 16

Interwar Bomber Ranges and Dates

Night or Heavy Bombers

Type	Spec. No.	Date in	Range	Crew	Buy
Aldershot	2/20	5/24	6 Hours	3	1 Sqn
Virginia	1/21	12/24	985	4	126
Hyderabad	31/21	12/25	500	4	45
Sidestrand	9/24	4/28	500	4	18
Hinaidi	13/25	10/29	850	4	33
Heyford	19/27	11/33	920	4	122
Overstrand	29/32	3/34	545	5	24
Hendon	14/27	12/36	1,360	5	14
Harrow	29/35	5/37	1,250	5	100
Whitley	3/34	/37	1,250	5	1,600
Wellington	9/32	/37	1,200	6	1,461
Hampden	9/32	8/38	1,200	4	1,270

Light or Day Bombers

Type	Spec. No.	Date in	Range	Buy
Fawn	5/21	3/24	650	77
Fox	21/25	8/26	500	28
Horsley	26/23	11/26	10 hours	119
IIIF	19/24	26	400	622
Hart	12/21	12/26	470	983
Gordon	18/30	4/31	600	160
Hind	18/30	11/35	430	527
Blenheim	28/35	1/37	1,125	5,000+
Wellesley	4/31	4/37	1,110	176
Battle	27/32	5/37	1,050	2,419

The Spec. numbers incorporate the year in which the Air Staff first asked for tenders to build the aeroplanes, and therefore the year in which the Air Staff was first thinking of how to use an aeroplane of this range, speed and load. This, for the purposes of this narrative, is the most important date. The Fox, Blenheim, and Wellesley were Private Ventures, that is built in hopes of a military sale before the RAF thought of buying them: in these cases the Spec numbers show the year in which the military version was contracted for. The same applies to the Anson, first built as a passenger aircraft for Imperial Airways.

The ranges given here are those claimed by the makers for combat range with a full load. With no load and with special tankage much greater ranges could be achieved, as with the Wellesleys of the Long Range Development Unit which managed one flight of 7,162 miles. The change in 1932 to standard ranges for both Light and Heavy bombers is obvious.

The Heavy bombers up to the mid-thirties were built for two pilots in one cockpit, one of whom did the navigating and bomb aiming. Pilots were hard to recruit and expensive to train both in resources and in time. To economize in pilots the NCO Observers, and later the Short Service Officer Observers, were brought in. As the war went on, an economy version of the Pilot was introduced to do second pilot work, and called the Flight Engineer. These men, usually NCOs, would if and when commissioned be carried on the General Duties List and not on that of the Technical Branch when it was formed. They were aircrew, and not ground or maintenance crew, and certainly not riding mechanics.

The Battle came in two fits, as conventional or as dive bomber. The spec in the dive bomber fit is uncannily a parallel to the Stuka, but with better range. Both the Battle and then the Stuka found that they were dreadfully vulnerable to a well sited and controlled anti-aircraft defence or to fighters when unescorted. The heavy losses for the Battle in 1940 led to its being withdrawn from operations. But: if the Germans had invaded using barges with jury-rigged anti-aircraft guns, and with their fighters kept busy upstairs then we might be now talking about the wonderful Bomber Command and the gallant Fairey Battles which won the Battle of Britain.

Table 17

Organization in the UK in September 1938

Bomber Command

1 Bomber Group
2 Bomber Group
3 Bomber Group
4 Bomber Group
5 Bomber Group
6 Auxiliary Group

Fighter Command:

11 Fighter Group
12 Fighter Group
22 Army Co-operation Group

Coastal Command
16 Reconnaissance Group
17 Training Group

Training Command:
23 Flying Training Group
24 Ground Training Group
25 Armament Training Group
26 Elementary and Reserve Flying Training Group

Balloon Command
30 Balloon Barrage Group

Two more Fighter Groups, 10 and 13, were formed before the war started. Two more Bomber Groups were formed early in the war, one to take in the Dominion Squadrons and the other to command the Continental Squadrons formed or reformed in the UK.

Table 18

The Group Pool Squadrons

All the Group Pool Squadrons were either formally disbanded or merged with the Operational Training Units already formed on their stations during April 1940, and were reformed at various dates as follows. Some were again disbanded and then reformed again later; this illustrates the ambiguity of the Squadron concept.

7B   reformed 8/40 with Stirlings.
35B   revived 11/40 with Halifax.
52B   title given 8/42 to MU at Habbaniya being uprated to Blenheim Squadron.
63B   reformed 6/42 as 63AC.
75B   title given to New Zealand Wellington Flight.
76B   reformed 5/41 with Halifax
90B   reformed 5/41 with Fortresses
97B   reformed 2/41 with Manchesters.
104   reformed 4/41 with Wellingtons at Malta.
108   reformed 8/41 with Wellingtons in Egypt.
148   reformed 12/40 with Wellingtons at Luqa.
166   reformed 1/43 with Wellingtons.
207   reformed 12/40 with Manchesters.

Table 19

Airmen's Trade Groups and rates of pay, 1919

Trade group I, technical: Blacksmith, Carpenter (Boat Builder), Carpenter (Motor Body Builder), Carpenter (Propeller Maker), Carpenter (Rigger), Coppersmith, Draughtsman (normally boys only), Electrician, (Compass setter and Repairer), Fitter and subsidiary combined trades (Fitters/aeroengine, general, MT, constructional, driver petrol or steam, motor boat, armourer, motor boat coxswain, millwright, jig and tool maker), Instrument Maker and Camera Repairer, Moulder, Pattern Maker, Turner, Wireless Operator (mechanic).

Group II Technical: Acetylene Welder, Balloon Basket Maker, Camera Repairer, Electrician, Machinist, Carpenter, Coach Painter, Photographer, Rigger (aero), Rigger (airship), Sheet Metal Worker, Tinsmith.

Group III Technical: Driver (Petrol), Motor Boat Coxswain, Driver (steam), Driver (winch), Motor Cyclist, Cook and Butcher, Shoemaker, Tailor, Musician, Hydrogen Worker, Vulcanizer, Motor Boat Crew, Stoker, Fabric Worker, Upholsterer.

Group IV Administrative: Clerk (general), Clerk (pay), Clerk (stores), Clerk (Q).

Group V Non-technical: Aircraft Hand (Batman), Aircraft Hand (G), Aircraft Hand (PTI), Aircraft Hand (Gunnery instructor).

Pay Scales, in shillings per day

Rank	Group I	Group IV	Group V
S/Major I	14/-	11/-	10/-
S/Major II	13/-	10/-	9/-
F/Sgt	11/6	8/6	8/-
Sgt	9/6	7/-	6/6
Cpl	7/9	5/10	5/-
LAC	5/6	4/6	4/-
AC1	4/6	4/-	3/4
AC2	4/-	3/6	3/-
Boy	1/6	1/6	1/6

In pay, Group II came between Groups I and IV, Group III between Groups IV and V. T.E. Lawrence spent his career as Aircraftman Shaw in a Group V trade, i.e. unskilled and the most poorly paid. It was possible to re-engage to complete twenty-four years service and qualify for a pension.

The list of trades shows clearly the technical structure of the servicing system and of the machines to be serviced.

Table 20

Airman Trade Groups and Pay Scales, 1934

Trade Group I: Fitter, Rigger, Fitter/Armourer, Fitter/Torpedo, Hydrogen Worker Class 1, Instrument Maker, Machine Tool Setter, Metal Worker, Wireless Operator/Mechanic.
Trade Group II: Armoured Car Crew, Armourer, Carpenter, Photographer, Rigger/Airship, Wireless Operator.
Trade Group III: Cook and Butcher, Fabric Worker, Hydrogen Worker Class II, Motorboat crew, Store keeper.
Trade Group IV: Clerk General Duties, Clerk Pay Accounts, Clerk Stores Accounts.
Trade Group V: Aircraft Hand General Duties, Physical Training Instructor, Police, Driver Petrol, Hospital Orderly, Musician, Apprentice, Boy Entrant, Skilled Entrant.

Pay Rates per annum
Trade Group

Rank	I	IV	V
AC2	£63	£57	£50
AC1	£77	£68	£54
LAC	£100	£82	£66
Cpl	£136	£100	£82
Sgt	£173	£127	£109
F/Sgt	£209	£155	£136
WO2	£237	£191	£182
WO1	£255	£209	£200

All these pay rates are the bottom of the scale in each case for the rank and trade, and would increase with years of service. Airmen also got accommodation and rations, or an allowance in lieu when detached or living out. There were also marriage and child allowances. Pay rates were in descending order for Group I (Technical), II (Technical), IV (Administrative), Group III (Technical), Group V (Non-technical).

NCO Pilots got the pay of their rank (Sgt or F/Sgt) and trade group, with an extra allowance of £54 pa for Sergeants and £64 pa for Flight Sergeants. Qualified Air Gunners, ground tradesmen who might be called away from normal work to fly, received an extra sixpence a day, as well as crew pay of two shillings a day while employed on the authorized establishment of a squadron for service in aircraft.

In looking at these pay scales, it should be remembered that during the 1930s pay of under £2 10s a week (£2. 50) was normal for a semi-skilled man in heavy industry, with no extra allowances for wives or children, or rent or rations. The pay for full-time Air Raid Wardens or Auxiliary Firemen on the outbreak of war was set at £3.00 per week: this was high enough to cause by attraction a severe shortage of skilled men (that is men who had carried out full apprenticeships) in essential war industries in the areas of traditional high unemployment, which could not compete either in wages or in security.

Table 21

Officer Pay Scales 1934

	Per Annum	
Air Chief Marshal	£2117	
Air Marshal	£1812 16s 8d	
Air Vice-Marshal	£1624 5s 0d	
Air Commodore	£1046 6s 8d	
Group Captain	£894 5s 0d	to £1040 5s 0d
Wing Commander	£650 18s 4d	to £842 2s 6d
Squadron Leader	£553 11s 8d	to £584
Flight Lieutenant	£442 15s 6d	to £456 5s 0d
Flying Officer	£325 9s 2d	to £374 2s 6d
Pilot Officer	£258 10s 10d	

It should be noted that the pay scales of all officers of the three services, and of soldiers, airmen and Royal Marines, and of all Civil Servants, were reckoned by reference to pay scales laid down in 1919 at the height of the wartime inflation. They were adjusted each year in accordance with the cost of living index; as the period 1919 to 1935 was a period of constant deflation, these scales were always adjusted *downwards*. This system was discounted from 1 April 1934.

Naval Ratings, being of a different legal status, were paid on steady scales which were not subject to these adjustments. At the time of the Invergordon Mutiny, the Government decided to apply all these adjustments to Ratings' pay in one go. The mutineers received little sympathy from other servicemen, and especially from their own officers, because they knew that their own pay went down every year.

Table 22

Civil Service Pay Scales, 1934

Grade	Per Annum
Deputy Secretary	£2,200
Principal Assistant Secretary	£1,200 to £1,500
Assistant Secretary	£1,000 to £1,200
Principal	£700 to £900
Assistant Principal	£200 to £500
Higher Clerical Officer	£300 to £450
Clerical Officer	£60 to £250
Shorthand Typist 28/- to	48/- per week
Typist 22/- to	36/- per week
Principal Scientific Officer	£650 to £750
Senior Scientific Officer	£560 to £600
Scientific Officer	£250 to £450
Junior Scientific Officer	£175 to £235
Civil Engineers	£450 to £621
Surveyors	£420 to £620

Note that the Executive and Experimental Officer classes do not yet exist, nor are any pay scales given in these estimates for the Technical Officer Class. No pay scale is given for the Permanent Secretary to the Ministry. In August 1936, the then PS was dismissed from the Civil Service, because on seeing his retirement approaching, he had suggested that he might be appointed a Goverment Director of Imperial Airways, a Quango. Such behaviour would be commonplace today but was then considered as an offence against Civil Service standards.

As late as 1949, a starting salary of £350 a year was normal for new University graduates in London.

Table 23

RAE Farnborough – Senior Staff Pay Scales and Manning, 1934

	From	To	Number
Chief Superintendent	£1,000	£1,200	1
Superintendent of Scientific Research	£800	£1,000	1
Principal Scientific Officer	£650	£750	5
Senior Scientific Officer	£500	£600	10
Scientific Officer	£250	£450	36
Junior Scientific Officer	£175	£235	17
Superintendent of Technical Development	£800	£1,000	1
Principal Technical Officer	£650	£750	4
Senior Technical Officer	£500	£600	6
Technical Officer	£250	£450	40
Junior Technical Officer	£175	£235	21

Assistant and Laboratory Assistants were paid on a variety of scales, allowing progress to a maximum of £260 pa. Women, where employed, and then normally only in the lowest grades, were paid less, and lost their jobs on marriage. These are provincial rates, not London scales.

Table 24

Officer and Aircrew Pay Scales, 1938

Air Chief Marshal	£2,311
Air Marshal	£1,983
Air Vice Marshal	£1,651
Air Commodore	£1,064
Group Captain	£958
Wing Commander	£660
Squadron Leader	£562
Flight Lieutenant	£428
Flying Officer	£331
Pilot Officer	£264
Acting Pilot Officer	£215

These are basic rates, as given on promotion, but pay went up by yearly increments to something just below the bottom rate of the rank above, as shown in earlier tables. Flying Pay is not included. These scales show a small advance on the scales of 1934.

Flight Sergeant Pilot	£273
Sergeant Pilot	£226
Sergeant Observer	£200
Corporal Observer	£164

These airman aircrew scales are now rates for the job, and independent of the NCOs original trade group. At all ranks, accommodation and rations or payment in lieu were drawn, and for Airmen and Officers over the prescribed ages, marriage allowance was paid.

Table 25

Scientific and Technical Pay Scales, RAE Farnborough, 1938

	From	To per annum
Chief Superintendent	£1,450	
Scientific and Technical Superintendent of Scientific Research or of Technical Development	£1,050	£1,150
Assistant Superintendent	£1,050	£1,150
Principal Scientific or Technical Officer	£850	£1,050
Senior Scientific or Technical Officer	£680	£800
Scientific Officer	£400	£680
Junior Scientific Officer	£175	£347
Technical Officer	£275	£580
Assistant I	£400	£575
Assistant II	£315	£385
Assistant III	£170	£310

Laboratory Assistant: up to £4 5s 0d per week

Civil Servants were not paid marriage allowance or flying pay, and were not entitled to accommodation or rations or to payments in lieu. This made them cheaper to employ than serving officers or airmen.

Note: at this date, the post of Assistant Superintendent of Technical Development was filled by a Squadron Leader RAF on service rates of pay.

Table 26

Civilian Strength,
Royal Aircraft Establishment, 1938–9

Grade	March 1938	Sept 1938	March 1939	Sept 1939
Superintendent Grades	6	6	6	9
Principal SO	7	8	11	8
Senior SO	15	15	15	14
Scientific Officer	52	53	53	54
Principal TO	7	7	7	8
Senior TO	29	24	26	30
Technical Officer	84	94	102	102

These were the posts which the Pilots with the 'e' annotation wished to see given to them.

Index

Abyssinia, 128, 155, 206
Aden, 88, 95, 96, 99, 206, 220
Admiralty, 42, 53, 55, 60, 66, 72, 73, 85, 93, 103, 168; Air Ministry and, 119ff, 133
aerial reconnaissance, 40, 51, 55, 60; see also balloon reconnaissance
Air Board, 58, 61: First, 61, 65, 69; Second, 61, 65, 69
Air Council, 65, 89, 158, 165, 181
Air Exercises (1933), 11, 161-2, 227
Air Force Estimates, 18, 84-5, 86, 89-90, 96-8, 99, 168, 178, 187, 188, 191-3, 195, 196, 223, 235
Air Force Lists, 18, 84-7, 96-9, 101, 103, 125, 130, 133, 143-4, 149, 156, 158-9, 173, 174, 192, 195, 208, 213, 214, 217-18, 221, 223, 233-5
Air Ministry, 50, 62-3, 65, 69, 79, 80, 81, 85, 94-5, 103, 114-16, 170, 191; Admiralty and, 119ff
Air Staff, 11, 62, 73, 151, 156, 158, 168, 184, 191, 193, 202, 204-5, 209-11, 221, 223, 226-8, 230; Manning Plans, 11, 184-6, 223, 226, 235-6, 237; Task Charts, 11, 150-1, 185, 223, 226
Air Training Corps (formerly Air Defence Cadet Corps), 238
airships, 35, 36, 43, 52, 192: Dirigible No. 1, 37; R100, R101, 90; Naval Airships 3 & 4, 119, 137
Army, 10, 20-1, 39, 40, 43, 64, 84, 89, 93, 101, 103, 131, 133, 142, 147, 148, 151, 161, 172, 181-3, 200, 201, 209, 221, 226, 227, 231, 237;
Army Education Corps, 150; Board of Ordnance, 20-1; Cttee of Imperial Defence created (later Imperial General Staff), 32-3, 38; Esher Cttee and reforms, 31-3; Cardwell's reforms, 22-4, 30; County Regiments, 23, 30, 31, 64, 77, 93, 98, 146; First World War, 45-67, 136; General Staff, 29-30, 32, 48-9, 50; mess structure, 145-6, 147; Militia, 22, 25, 31; Officers' Training Corps, 136-7, 211; Officer Training School (France), 111-12; officer recruitment, selection, 20-2; 1905, 136; purchase, 22, 33, 139; recruitment and training, other ranks, 106-8, 111-13; Royal Military Academy (Sandhurst), 112, 137, 139, 173; Staff College (Camberley), 27, 30, 33, 37-8, 173; Staff College, Quetta, 37, 54; uniform, 34, 183-4; Volunteers, 25, 31; see also Royal Flying Corps, Military Wing
Army Lists, 119
Atcherley, Wing Commander R.T., 232
Atlantic, Battle of, 231
Austria, 45, 77, 154, 220-1
Australia, 25, 80, 82

Bailhache, Mr Justice, 61
Baldwin, Air Vice-Marshal, 141; Cttee, 61
Baldwin, Stanley, 155, 160-1, 184, 234
balloon reconnaissance, 33-5, 60:
American Civil War, 33-4: Boer War, 25; Kite (stabilized) balloons, 34-5, 51,

80; Naval, 52; vulnerability, 34-5
Balloon School and Factory, Farnborough, 33, 35, 37
Baring, Hon. Maurice, 48-9, 54, 57-9, 77, 79, 159, 170, 182
Basra, 99, 220
Battle of Britain, 10, 12, 16, 17, 19, 94, 154, 168, 175, 230, 235, 236
Beatty, Sir David, 73, 90, 119
Belgium, 45-6, 60, 64, 67, 78, 233
Billing, Pemberton, 53-4, 59
Blériot, Louis, 40
British Expeditionary Force, 46-8, 54-7, 67, 208
Broke-Smith, Lieut., 35
Buller, Redvers, 24, 29, 30, 58
Burma, 80, 229

Calabar (Nigeria), 26, 112
Campbell-Bannerman, Sir Henry, 32
Canada, 25, 72, 82
Cape Colony, 24, 25, 26
Capper, John, 35-6, 37, 80, 192, 226
Capper, Thompson, 37, 54
Cardwell, Sir Edward, 20, 22-4
Chamberlain, Austen, 122
Chamberlain, Sir Neville, 155-6, 234
Churchill, Winston, 24, 60, 67, 78, 234
civil aviation, 67, 78, 79, 81, 90; *see also Air Estimates, Air Force Lists*
Clark-Hall, R.H., 180
Cody, S.F., 35, 37, 39, 192
Courtney, Sir Christopher, 165
Cowdray, Lord, 61
Crimean War, 31, 39
Cross, K.B., 232
Curzon, Lord, 61
Czechoslovakia, 154-5, 207, 220; Army, 233

Dardanelles campaign, 55, 61
Denmark, 28, 230, 233, 235
Dowding, Hugh, 41-2, 49, 156, 165, 226, 230, 236

Edward VII, King, 31, 32
Egypt, 70, 83, 90, 96, 99, 127, 144, 206, 207, 213, 220

Ellington, Edward, 156, 162, 164, 226
Embry, Basil, 216, 218
Esher, Lord, 20, 31-3, 47

First World War, 10, 45-67, 93, 98, 101, 111-12, 114-15, 119, 136, 148, 153, 155, 160, 182, 193, 226, 236
Fleet Air Arm, 119, 122, 127, 129-30
Flight, 18, 83, 99, 110, 113, 115, 139, 162, 163, 216
Foch, Ferdinand, 46, 66, 77
France, the French, 10, 28, 29, 30, 32, 34, 38, 45-67 *passim*, 70, 75, 76, 80, 89, 101, 103, 111, 146, 163, 200, 226, 230, 232, 236; aircraft carrier *Bearn*, 121; Air Force, 233
French, John, 24-5, 46-7, 56

Gallipoli, 55-6
General Staff (British), 46, 48-50; concept of, 27-8, 29, 47; German, 29, 30, 47-8
George V, King, 47, 64, 69
German Air Force, 9, 61, 154, 161, 163, 177, 222, 228, 230-1
German Army, 10, 45-67 *passim*, 137, 154: balloon observation and, 34; General Staff, 45, 47, 232-3; manoeuvres (1936), 200-1, 221
German Navy, 53, 128-9, 147, 162, 192, 227; *Bismark*, 131; High Seas Fleet, 53, 63, 120; *Gneisenau, Prinz Eugen, Scharnhorst, Tirpitz*, 232
Germany, 31, 32, 37, 53, 74, 77, 135, 151, 153-5, 158, 200, 235

Haig, Douglas, 24, 47, 54, 56-7, 63, 64-5, 67, 75, 77, 78
Haldane, Viscount, 20, 32
Hampton, H.N., 137
Harris, Arthur, 236
Henderson, David, 10, 24, 38, 43, 48-9, 54-5, 56, 59, 62-3, 73, 76, 79, 89, 226
Hitler, Adolf, 11, 18, 153-4
Holland, 53, 233
home defence, 62, 71; *see also* under RAF, Squadrons

Independent Air Force, 11, 70, 73-8, 137,

161, 164, 217, 226
India, 20, 23-4, 26-7, 33, 37, 70, 80, 83, 90, 96, 99, 102, 169, 220, 229; Army, 37
Inter-allied Aviation Committee, 76-7
Iraq, 96, 164, 220
Ireland, 23, 32, 41, 70-1, 102
Italy, 16, 66, 70, 128, 155, 233

Japan, 16, 25, 103, 121, 122-4, 163, 207, 229
Jellicoe, Sir John, 120
Jillings, D.S., 50, 135-6
Journal of the Royal Aeronautical Society, 18

King-Hall, Stephen, 236
Kirby, H.H., 24, 42, 149, 192, 195
Kitchener, Horatio, 24-5, 47, 50, 61

Ladysmith, siege of (1899-1900), 31
Law, Bonar, 122
Lawrence, T.E. (Aircraftman Shaw), 105, 108-9, 113-14, 137-8, 175
League of Nations, 154, 155
Lloyd George, David, 10, 14, 25, 32, 61-2, 63-6 *passim*, 74, 80, 161, 226; Khaki Election and, 67
Lutyens, Sir Edwin, 170; 'Lutyens' Stations', 170-3 *passim*

MacDonald, Ramsay, 155, 161, 234
Malta, 97, 99, 128, 155, 220
Marne, Battle of the (1918), 48
Matapan, Cape, Battle of (1941), 131
Mediterranean area, 70, 73, 82, 96, 120-1, 125, 127, 128, 155, 220
Mesopotamia, 70, 71, 77, 83, 96, 99, 217
Midway, Battle of (1942), 123-4
Military Aeroplane Competition, 40
Munich Crisis (Sept. 1938), 17-18, 155, 207, 220, 223-4, 236

Natal, 24, 29, 58
Naval Annual, 18
Navy Lists, 88, 119, 126, 127, 128, 148
Nazi Party, 153-4
Newall, Cyril, 75-6, 156, 164-5, 226
New Zealand, 25, 82

Nigeria, 26, 41
Northcliffe, Lord, 62, 63, 65
Norway, 230, 231, 233

Paine, Godfrey, 42, 73
Palestine, 63, 70, 71, 77, 96, 220
Paris, siege of (1871), 31, 33, 48
Pollard, Captain Guy, 143
Portal, Charles, 140, 144

RAF Quarterly, 18, 189-90, 200
Roberts, Field-Marshal Sir Frederick, 24-5
Robertson, William, 24, 27, 38, 46-7, 54, 56, 75, 78, 111-12
Roe, A.V., 59
Rothermere, Lord, 62, 65, 73, 74, 75, 184
Royal Aero Club Certificate of Proficiency in Flying, 40-1, 42
Royal Aircraft Factory, Farnborough, 39-40; *see also* Balloon School and Factory
Royal Air Force, creation of (1917), 62-7: squadron dispositions, 69-71, 73; post-war disbandment (1918), 78; training, 71-3, 80; Gosport system, 72, 140; Sykes Memorandum on, 80-4, 92-4 *passim*, 187, 226, 238; Trenchard Memorandum on, 80-4, 89, 93, 96, 106; Geddes recommendations, 96; airfields, 167-8; *see also* stations
aircraft design, purchase, 114-17; armament, 116-17, 178
Apprentice Clerks (1934), 188
Apprentice Training, 106-7, 109-14; post 1918 (officers and men), 137
Bomber Offensives, 10, 12, 14, 211, 238
Boy Entrants scheme (1935), 188-9
commissions, posts, mess system, 145-7, 150, 170-3
education service, 150
Empire Air Training Scheme, 213
flying training, 112, 160, 178: airman pilots, 112-13, 133, 157; officers, 157
Ground Branch officers, selection, 199-202
Group Selection procedures, 201-2; pay, 90, 192; *see also* Tables pilot recruitment and training, 133-5, 167-8, 184, 213-14; Acting PO rank, 218

ranking and promotion (officers), 86-9, 142-219, (men), 113
recruiting (post-1922), 142-5
radar (Control and Reporting), 163-4, 165, 168, 170, 199, 202, 210, 222-3, 228
radiolocation, 227
rebuilding and expansion (post-1933), 167-73, 201
reserves, 79, 142, 143, 144, 158, 159-60; Reserve of Air Force officers, 79, 159, 174; *see also* Volunteer Reserve
Royal Navy and, 124-31 *passim*
Second World War, 227-38; preparation for, 156-66: airfields, 223 (*see also* stations); air gunners, 156, 216, 217; Acting PO rank, 218; Empire, 220; manning crisis, 218-20; navigators, 216-18, 228; observers, 217-18; re-arming (post 1937), 223-4, 234; Schemes 'A', 'C', 'F', 'M', 165-6
service and maintenance of aircraft
short-service commissions, entry, 108, 133, 137, 142-3, 145, 158, 159, 171, 175, 178, 210, 211, 212
trades, 108-10, 113, 145, 167, 185-9; supervision, 188-9
Volunteer Reserve, 198-9, 211-13: Balloon Branch, 199; Admin. and Special Duties Branch, 199

INDIVIDUAL ESTABLISHMENTS:
  Aircraft and Armament School (Martlesham), 95, 159, 192
  Aircraft Experimental Establishment (Martlesham), 102
  Air Pilotage School, 149, 217
  Airship School (Cranwell), 72
  Anti-Aircraft Co-operation Flight (Biggin Hill), 134, 181
  Apprentice School (Cosford), 179
  Artillery and Gunnery Schools (Uxbridge, Eastchurch), 73, 102, 159, 186
  Balloonist Schools (Larkhill, Lydd), 72, 180
  Boy Training (Eastchurch), 72
  Cadet Brigade (Shorncliffe), 72
  Central Flying School (Upavon), 133-4, 210-15 *passim*
  Electrical & Wireless Schools (Flowerdown, Penshurst), 73, 102
  Initial Training School, 137, 212, 213
  Home Aircraft Depot (Henlow), 71, 78, 134, 144-5, 149, 159, 180, 188, 189-90, 193, 196
  flying training schools, 11, 38, 127, 133-4, 138, 143-4, 157, 160, 169, 180, 186, 195, 207, 212-13, 217
  No. 1 School Technical Training (Men) (Halton), 72-3, 109-11, 113, 117, 134, 144, 176, 178-9, 186-7, 188, 191
  No. 3 School Technical Training (Manston), 102, 114, 180, 185, 186-7
  Royal Aircraft Establishment (Farnborough), 102, 144, 190, 192
  Royal Air Force College, Cranwell, 108-11 *passim*, 114, 133-41 *passim*, 144, 145, 160, 173, 176, 191, 192, 201, 211
  School of Army Co-operation (Old Sarum), 102, 180, 215
  School of Air Navigation, 218
  School of Free Ballooning (Hurlingham), 72
  School of Photography (Farnborough), 73, 102, 159, 180
  Signals Co-operation Flight (Biggin Hill), 102
  Staff College (Andover), 102, 150, 159, 181
  Wireless Apprentice School (Cranwell), 111, 114, 117

ORGANIZATION
  Commands, 193, 223:
  Air Defence of GB, 180-1
  Bomber, 181, 207, 208-9, 227-9, 232
  Coastal, 127, 181, 231
  Fighter, 9, 165, 181, 222, 223, 229-30
  Maintenance, 181, 183
  Training, 180, 181
  Areas, 101-2, 103, 179-81
  Branches:
  Accounts, 134, 145, 174, 196-7
  GD, 88, 133-5, 144, 149, 190, 193, 194-5, 196, 197

Medical and Dental, 134, 149, 197
Stores (and Depots), 102, 134, 144-5, 149
Technical, 194-5, 199
Groups, 100-1, 134, 168, 180-1, 193, 223, 230
Wings, 101, 134, 144, 215
Squadrons, (RFC) 43, 49, 51 (RAF), 69-70, 88, 92-104, 141, 144, 149, 151, 159-60, 161, 163, 164, 168-9, 171, 176-81 *passim*, 187-9, 193, 204, 208-24 *passim*, 226; *see also* Stations
balloon, 101
on aircraft carriers, 119, 125, 126, 179, 231
Army Co-operation, 95, 97, 99, 103, 142, 156-7, 178, 205-8, 220, 223
bomber, 60, 61, 70-1, 75, 76-8, 82, 88, 94-9 *passim*, 101, 103, 115-17, 131, 156, 157-8, 161-5, 168, 174, 178, 179, 181, 205-9, 216, 218, 220, 223, 226-30, 231
cadre, 81-2, 97-8, 99, 134, 160, 177, 180, 205-6, 221
communications, 206-7
fighters, 60, 70-1, 82, 88, 94-101 *passim*, 103, 115, 116-17, 126-7, 131, 156-7, 161-5, 178, 181, 205-9, 216, 218, 229-30
catapult, 129-30
flying boats, 95, 99, 156-7, 177, 179, 205-7, 215, 217, 220, 231
'Group Pool', 208-9, 216
home defence, 96-8, 99, 101, 103, 164, 180, 182
observation, 70-1, 101
photography, 96, 102
reconnaissance, 70-1, 82, 95, 101, 126-7, 131, 162, 206, 208, 216, 223, 231
reserves, 97-9, 160
training, 71, 72, 78, 101, 177-8
University Air Squadrons, 181, 211, 215
Stations, 100-1, 134, 141, 147, 149, 168-82: administration, 175-6; aircraft, 174, building, 170-3; functions of typical, 173-7; manning, 173-5; numbers, 177-81; rearmament, 176-7; servicing, spares, 176-7; system of messing, 170-2; training, 176, 177

Royal Artillery, 10, 11, 20, 41, 58, 93, 138
Royal Engineers, 10, 20, 34, 37, 42, 51
Royal Flying Corps, 38ff, 69-70, 71, 78, 96, 105, 112, 136, 199; constitution of, 38, 62, 69; aerial reconnaissance, 40; aircraft trials, 39-40; armament, 50-1; Bailhache Inquiry, 61; First World War, 54-6; HQ, 49, 57-9, 75-6; Military Wing, 38-40, 41-3, 48-51, 52, 70, 72; Naval Wing, 38, 42-3, 51-4, 180; pilot training school, 38; raids by, 74; reserve, 38, 40, 41; squadrons established, 38-9, 208; uniform, 39
Royal Marines, 53, 55, 202
Royal Military College, Sandhurst, 20, 22-3, 26
Royal Naval Air Service, 53, 55, 59, 60-1, 70, 74-5, 77, 78, 88, 95, 119, 182, 217
Royal Naval College, Dartmouth, 20, 137
Royal Navy, 10, 21, 59, 84, 90, 93-4, 103, 147, 148, 181, 193-4; aircraft carriers, development, 11, 100, 103, 119-31 *passim*, 134; apprentice training, 106; airship station (Farnborough), 119; balloons and, 52; bombing raids by, 60-1, 74; flying school, 42; home defence and, 60, 82, 128; mess system, 146-7; Naval Flying School (Eastchurch), 119; ranking and promotion, 124-5; RAF and, 124-31, 133, 142; reconnaissance and, 52, 120; short-service commissions, 127; squadron dispositions, 70-1ff, 97; submarines, 70, 78; *see also* Royal Flying Corps, Naval Wing

classes – *Arethusa* destroyers, 52; *County* cruisers, 129; *Royal Sovereign* battleships, 130
ships – *Argus*, 102, 121, 125, 127; *Ark Royal* (renamed *Pegasus* 1935), 120, 122, 125, 127, 129, 130; *Courageous*, 122, 126, 127; *Eagle*, 102, 121, 126, 127, 130; *Furious*, 121, 122, 126, 127,

130; *Glorious*, 122, 126, 127, 130; *Hermes*, 102, 121, 126, 127, 130; *King George V*, 130; *Pegasus*, 120, 125; *Vanguard*, 122

RUSI Journal, 18

Russia, 45, 49, 64, 67, 90, 154, 155, 163

Salmond, Geoffrey, 49, 151, 164, 180
Salmond, John, 49, 54, 67, 75, 76, 151, 156, 162, 164
Samson, 42, 55-6, 60, 119
Second Boer War, 24-7, 29, 30, 31, 34, 38, 45, 50, 53, 58, 96, 182, 227
Second World War, 9-11, 16, 52, 53, 73, 89, 120, 121, 139, 147, 150, 207, 236
Singapore, 99, 122, 205, 220
Slessor, John, 136-7
Smith-Barry, Robert, 72, 136, 140, 210
Smith-Dorrien, Horace, 47
Smuts, Jan Christian, 25, 62-3, 226
Somme, Battle of the, 74, 112, 211, 222
Sopwith, T.M., 59
South Africa, 24-6 *passim*, 37, 38, 41, 42, 48, 62, 67, 82
Spain, 145, 154, 222, 233, 235
Spanish Civil War, 128
Swinton, E.D. ('Ole Luke Oie'), 51, 53, 66, 79, 80
Sykes, Frederick, 10, 24-6, 27, 33, 37, 38, 56, 80, 137, 144, 164, 170, 183, 186, 202, 219, 221; background, 24; proposals to set up RFC (1911-12); 38-9, 43; Military Wing RFC (1912-14), 43, 48-51, 54-5; Dardanelles (1915), 55-6; Colchester, War Office, tanks, WAAC (1916), 56, 66; Chief of Air Staff (1917), 65-7, 69, 73-7 *passim*; Memorandum (1918), 80-4, 92-4 *passim*, 187, 226, 238; Controller-General Civil Aviation (1919), 67, 78-80

tanks, 66, 79, 109

Tank Corps, 80, 108
Taranto, Battle of (1940), 130
Tedder, Arthur, 236
Thorne, General Sir Andrew, 200-1
Transvaal, 24, 25
Trenchard, Hugh, 11, 14, 109, 140, 142, 164, 177, 190, 193, 194, 198, 202, 217, 219; background, 20; Army service, 20-6; Military Wing RFC (1912), 41-3, 49, 54-5, 56; Chief Military Wing (1915), 55-7; visits to squadrons, 58-9, 115; first Chief of Air Staff (1917), 65; resigns (1918), 65, 184; Independent Air Force and, 73-7; reappointed Chief of Air Staff (1919), 67, 73-5, 78-90 *passim*, 99, 106, 109, 110-17 *passim*, 124, 133, 135, 137-42 *passim*, 147, 149, 150-1, 158, 181; Memorandum (1919), 80-4, 89, 93, 96, 106; manning plan, 185-6

United States, 25, 36, 103, 154; Air Force Academy, 138; Army, 24, 67, 148, 149, 163, 199, 207; Army Air Force, 163, 228; Navy, 124, 130

Versailles, Treaty of, 78, 79, 154

War Office, 10, 36-8, 46, 48, 54, 60, 66, 79, 80, 85, 103, 136, 201
Weir, William, 65, 67, 73-6 *passim*, 78, 167, 226
Whittle, Frank, 193
Wilson, Henry, 24, 38, 46-7, 54-5, 56, 63, 66, 79
Women's Auxiliary Air Force, 198-9, 202
Women's Auxiliary Army Corps, 66, 202
Wright, Frank and Orville, 35-6, 51

Zeppelin, Count, 36
Zeppelin airships, 36-7, 52-4, 60-1, 74, 120; raids on UK, 61-3 *passim*